The String and Glue
of Our World

T0197960

The String and Glue of Our World

Understanding Composite Materials

NED PATTON

Foreword by Michael Santare

McFarland & Company, Inc., Publishers
Jefferson, North Carolina

ISBN (print) 978-1-4766-9133-6
ISBN (ebook) 978-1-4766-4994-8

LIBRARY OF CONGRESS AND BRITISH LIBRARY
CATALOGUING DATA ARE AVAILABLE

Library of Congress Control Number 2023028888

Front cover design by Candace Hignettå

Printed in the United States of America

*McFarland & Company, Inc., Publishers
Box 611, Jefferson, North Carolina 28640
www.mcfarlandpub.com*

Table of Contents

Foreword

by Michael Santare

Dr. Patton has been in the business of engineering for 41 years. He has focused on research and development of composites (among other things) for the last 38 of those. His career has taken him to private industry, government research, and back. I had the pleasure of meeting him when he went back to school to pursue his PhD in mechanical engineering at the University of Delaware in 1987, while he was working as a team leader at the U.S. Army's Ballistics Research Laboratory (now ARL).

What struck me most at the time was his clear sense of purpose and determination when he came into my office and announced that he planned to do his PhD research with me as his advisor. I was fresh out of graduate school and thought I knew plenty about how to conduct applied mechanics research. However, the next few years were a transformational time for me as I learned that advising students' research work involves learning as much from them as they from me. One of Ned's strong points has always been the ability to break complex issues up into smaller, easier-to-digest pieces and putting them back together to address the larger problem.

That's exactly what he has done for the reader in this book. He takes the daunting amount of information out there in the field of composite materials and makes it understandable, in a readable and relatable way. In the first chapter, Ned gives a brief history of composites. This historical approach becomes a unifying theme throughout the rest of book as he interweaves anecdotes about key people and important milestones to bring the rest of the narration to life. In the next few chapters, he explains the chemistry of the various resin and fiber materials commonly used, the "string and glue," as he puts it, and how they can be put together to make composite materials. He does this in a way that the reader with a high-school level background in chemistry can see the similarities and differences among constituents, leading to the relative advantages and drawbacks of choosing one over the other.

1

He then gets into the mechanical performance of composites, where the rubber meets the road, so to speak. In this part of the book, Ned introduces some of the basic concepts needed to describe and understand the mechanics of composites—stiffness, strength, strain, and stress. Again, he does this in a way that anyone with a high school education and who has attached a board to a rope and made a swing will understand. Building on those basic ideas, he describes the mechanical behavior of the fibers and how the fibers and resin (string and glue) work together to make a composite material with properties beyond those of either constituent. He goes on to show how these unique behaviors of composites can be employed by introducing some basic ideas about how to design and make useful things out of these materials. And, just as importantly, he lets the reader know how to avoid disaster by explaining how composites can fail if they aren't made or designed properly. All the while, he brings in pieces of the history of composites and uses anecdotes to make things clear for his audience.

In the last few chapters of the book, Dr. Patton gets into the business and education aspects of the composites industry. Using anecdotes about some of the seedier sides of the business, Ned provides a bit more history about how the composites industry became what it is today, and who had to pay who millions of dollars for what, and how to avoid going to jail—at least in the early days of the industry. He concludes with a broad list of universities that have good programs for composite material design, manufacturing, and mechanics. That list is extensive, and of course it includes all the larger and more well-known universities.

With this book, Dr. Patton brings his four decades of experience in design and analysis of composite materials to the science-literate reader in an engaging and accessible way. *The String and Glue of Our World* is an ideal introduction for people who are not necessarily trained in engineering but have some background knowledge in science and want to learn more about the field of composites. Overall, the book strikes a balance between the 30,000-foot overview and a specialized technical book on the subject. It presents an excellent introduction to composite materials for the science-literate reader who is considering getting involved in the field. It is also worth noting, that for the reader who knows something about the science and engineering behind composites, the book provides a delightful perspective and historical context.

Michael Santare earned his BS in mechanical engineering from Rensselaer Polytechnic Institute and his MS and PhD in theoretical and applied mechanics from Northwestern University. He is a professor of mechanical engineering, biomechanics and movement science and biomedical engineering at the University of Delaware where he has taught and developed a wide variety of courses in the general area of mechanics for the past 35 years. His research focusses on deformation and failure of complex materials systems, such as composites, biological tissues and ionomer membranes. He has authored or co-authored more than 100 research publications and has been invited to give numerous lectures and presentations nationally and internationally.

Preface

The String and Glue of Our World is a book about composite materials written from the perspective of someone who has been in this business for 40-plus years at this point. Unlike most books about composites, it is intended for a more general audience, so it is written in the first person and without much of the jargon that comprises similar books. It is not a textbook, nor is it a how-to book. Rather it is more of a general audience book that introduces composite materials in a friendly, approachable, and hopefully interesting and entertaining way. Using a *semantic tree of knowledge* approach, it starts with the basic history of composites and where the main ideas for these materials came from. Composites are much easier to understand if you start with the fundamentals and get the concept that most of the ideas that composites are based on are analogs of what we see in nature. Using examples from Mother Nature to explain the basics of composites makes them much easier to understand.

The book continues with a discussion of the relationship of composites to the periodic table of the elements and why the elements most important to composites are at the top of that table. The next set of branches on the semantic tree delve into the strings (fibers) that are used in composites and why and the glues (resins) that stick the strings together, and then I go on to describe how to make things using composites. Then there is a section on mechanics and application to design, failure and how to avoid it, computer-based tools for analysis and design of composites, some different types of composites, the business of composites, and finally where to get an education and find a job in this business.

I certainly hope you enjoy this book, and I invite anyone that is curious to learn more to contact me at my website http://nedpatton.com with questions. I'm here to answer whatever questions you want to ask and also to begin a dialog with interested folks who want to learn more about these wonderful materials.

Introduction

They call it the string and glue. Fiber-reinforced composites—or "fibre" if you speak the Queen's English. But since this book is being written in the United States, I'm sticking with fiber. The science, engineering, and art of making things out of composites lives where engineers (makers of things) meet up with the periodic table and they make friends with each other. You can too—even if you're not a PhD engineer or scientist. All you have to do is understand where the idea for composites came from, what part of the periodic table of the elements makes up the bulk of the matter in composites and why, and from that understand the idiosyncrasies of the fibers (strings) and resins (glue) when they come together to make magic happen. Just follow me on this journey into this fascinating subject and maybe learn how to make the magic happen for yourself—become a composite material aficionado.

This is a book about the science, engineering, history, and art of making things out of composites, written in a manner that is hopefully approachable, engaging, and entertaining, even for the casual reader. It is not intended to replace or even be a textbook on the subject—there are plenty of those. It is also not intended for the serious practitioner who has lots of experience with composites and needs a reference book, since there are lots of those as well. It is intended more for a general science- or engineering-curious audience that is interested in how things go together and work and why lots of things are made out of composites these days—including large pieces of the plane you just took to visit Grandma. In other words, I am calling out to all citizen scientists and makers out there who are interested in composites and want to learn more about these wonderful materials.

The book is arranged in what Elon Musk describes as the most important way to view knowledge—as a semantic tree. In that vein, I introduce first the history and motivation for the *why* of composites, then start with the building blocks of these materials: the elements themselves from the periodic table and why each of the elements I focus on are important.

This forms the trunk of the semantic tree of knowledge of composites. Then I build the branches of the tree with descriptions of the strings and glues, how they are put together, and who is putting them together. Of course, each of these three main branches has its own sub-branches. And then toward the end of the book I give you some ideas about how to get involved in this industry and make the magic happen for yourself.

I've been at this for more than 40 years now, a mechanical engineer with a PhD from the University of Delaware's Center for Composite Materials—one of the premier composites manufacturing technology centers in the United States. And I've been mostly in R&D and technology development my entire career and I love it. I'm usually the guy with the crazy idea about how to make something work. So far my successes have outnumbered my failures, so I have done OK. And since I've been at this for as long as I have, I have already made most of the mistakes and have the tire tracks on my back to prove it. This book comes from that perspective— from someone that has been there, done that, and can provide guidance to anyone who is just starting out with these wonderful materials.

I certainly hope you enjoy this book and get something from it, and just to make this real, I welcome anyone who has read the book or is just starting to read it to contact me with any questions or comments. I would love to start a dialogue with other interested people and see where this journey takes us.

Like they say, it takes a village.

Enjoy.

1

A Brief History of Composites

It all began some 13.8 billion years ago or so when there was a really large explosion and all matter and energy in the universe came into being. No? More seriously, all composite materials have an inspiration or analog, if you will, in Mother Nature. All that started with the Big Bang and its aftermath, which created all the known elements that man has found in nature—plus all the stars, galaxies, etc.—and many billions of years later, our solar system and Earth itself. And since all composites have some sort of analog or inspiration in Mother Nature, I thought I would start there.

To get started, let's get back to composites. What was the first completely man-made composite material, you ask? Concrete—one of the first man-made composites—came about when early Middle Eastern builders circa 6500 BC discovered that if they crushed and burned limestone, then ground it into a paste with water, it would bind sand and gravel together and make bigger, stronger rocks than did the limestone, gravel, and sand individually.[1] And they could shape this mixture into bricks and stack them to make structures. Or, as the Egyptians did circa 3000 BC, they could eliminate the gravel and use a slurry of sand, fired limestone, and water as a mortar to hold together large stones and bricks made of dried mud and straw. The Egyptian pyramids still exist today after 5000 years, so these early materials had a very long lifespan.

This is in fact the definition of composites. Mix two (or more) different materials together in the right way and in the right amounts and you get a new material that has more useful properties than each of the different materials do on their own. This is where we start building from the base of the trunk of the semantic tree—an understanding of what exactly a composite material is. Now let's go on to build the rest of the trunk and then the branches and leaves and see if we can get this thing to flower and bear fruit for us.

First, however, I need to define some terms that I use throughout the book to describe the characteristics of composites and why these materials are so important to us in our modern lives. The words I will use are strength, stiffness, stress, and strain, and all of them are used constantly by mechanical and structural engineers to describe the behavior of the materials that we use to make useful things. One example that almost everyone can understand is to think of a rope swing hanging from the branch of a tree. When you buy a rope to make your swing, you buy a rope that is strong enough to support the weight of anyone who might sit on the swing, even if they try things like "over the moon" on the swing. While that is pretty dangerous and I don't recommend it, when you buy your rope, it is wise to buy one that's quite a bit stronger than you need. Let's say that the swing is really intended for your kids, but you know that your younger brother who is 6'4" and weighs 200 pounds will probably also be on the swing. To be safe, you need a rope that will handle around 1000 pounds before it breaks. The number of pounds that it takes to break the rope when you pull on it is its tensile strength.

Now for stress and strain. If you make your rope swing with a simple board seat at the end of it, everyone knows that the heavier you are, the more the rope stretches when you sit on it. You also have to decide how far off the ground you need to put the board so that when someone swings on it they don't hit the ground. This is what is called strain in the rope. Engineers define strain as the amount that the rope stretches divided by the original length of the rope, so it is what is called a *dimensionless* number. This means that if your rope was originally 10 feet (120 inches) long, and when your brother sits on it, it stretches a foot (12 inches), the strain in the rope is 12/120 (or in feet 1/10) or 0.1 inches per inch (abbreviated as in/in). Now, what is stress, and how is it different than strength? Engineers define stress in something like the rope as how much force is applied to the rope divided by the cross-sectional area of the rope. This number, along with strain, are the two numbers that mechanical and structural engineers work with to design things so that they are not only useful but also safe to use. The dimensions of stress are force / area squared, or in English units, pounds per square inch. Let's say that your rope is 1" in diameter, and we already know that your brother weighs 200 pounds. The cross-sectional area of your rope is 0.785" (that's the radius squared times pi [π] so you don't have to remember your geometry), and the force in the rope when your overweight younger brother sits on it is 200 pounds, so the stress in the rope is 200 pounds / 0.785 or 255 pounds per square inch (abbreviated as psi).

Now on to stiffness. This concept is slightly more difficult to understand, but it is one that is fundamental to understanding why engineers use composites in the first place. The stiffness of a material in the parlance of engineers is also called the elastic modulus of the material or the Young's modulus of the material. It is a property of every material that behaves in what is called an *elastic* manner. If the rope goes back to its original length when your brother gets off the swing, then the stretch that happened when he sat on the swing was elastic stretching or elastic strain. The stiffness of the rope, or its Young's modulus, is defined by the stress in the rope when your brother sits on it divided by the strain that we calculated earlier when he first sat down on the rope, or 255 psi divided by 0.1 in/in or 2550 psi.

So that you can find them easily when you come upon them, here are the definitions:

- **Strength or Tensile Strength**—the force it takes to break the rope
- **Stress**—the force in the rope divided by its cross-sectional area
- **Strain**—the amount of stretch that the rope undergoes when your brother sits on it
- **Stiffness—elastic modulus or Young's modulus**—the stress in the rope divided by the strain in the rope when your brother sits on it

As an engineering material, rope is really not that stiff, since common metals have stiffnesses (Young's modulus) on the order of 10 to 30 million psi (abbreviated msi for million psi). Aluminum has a stiffness of 10 msi; steel, almost 30. Some graphite fibers can have a Young's modulus (stiffness) as high as 50 msi.

Now that we understand these basic concepts, we can go back to Mother Nature. Some of the best examples of Mother Nature–inspired composites are the modern—some say "advanced"—composites like carbon/epoxy and fiberglass. If you look at the cross-section of a straight-grained fir board you will see what a well-structured 3D composite looks like. This is the structure that man has been trying to copy with the development of modern advanced composites. The fibers in the fir board are made of cellulose, which is essentially the string that Mother Nature has devised to use as the strength and stiffness element of her composites. The glue is a material called lignin that is mixed with the pitch or the sap of the tree; it makes a very effective glue. The cellulose fibers are oriented in the direction that the tree needs strength and stiffness (mostly up and down), and the lignin binds these together and keeps the tree in one piece. The orientations of the cellulose fibers are not all aligned up and down the trunk of the tree so it has stiffness and strength both in the up/down axis of the trunk as well as across the trunk. The tree has to support its own branches, after all. Anyone who has ever split oak for firewood knows what I'm talking about here. It ain't easy—and that's because there are fibers that run not only in the direction the tree grows, but some also connect the inner layers of the trunk with the outer layers of the trunk. And, of course, the glue (lignin + a little pitch) in oak trees is a harder and

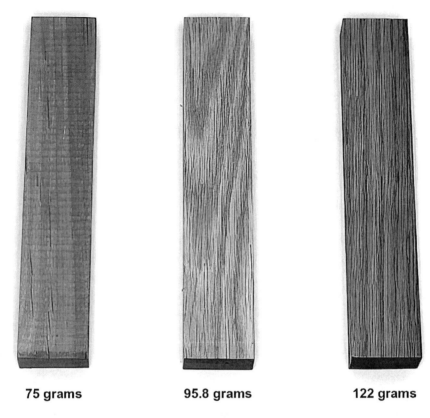

75 grams   **95.8 grams**   **122 grams**

Different woods—alder (left) and oak (middle and right) (photograph taken by author, enhanced by Candace Hignett).

more brittle material than the glue (lignin + more and different pitch) in pine and fir trees, so Mother Nature has a variety of different strings and glues at her disposal in the plant kingdom. A good example of a couple of different wood species and how the strings and glues are arranged is shown in the figure above.

　　If you look at this picture carefully and think a bit about the trees these samples came from, you will understand why I start with this analog. The two pieces on the right are both oak, so they should weigh the same, right? The reason they don't is that the one on the far right is from the heart of the oak tree, where the tree needs more compressive strength and compressive stiffness (gravity sucks, right?), whereas the one in the middle, while still being oak, is from farther out on the trunk of the tree where the tree needs more flexibility, but also needs to support the branches in the canopy. The center of the trunk has more dense packing of the heavier fibers as well as the incorporation of some metal ions like iron (hence the reddish color) to stiffen and strengthen the fibers and the glue. On the far left we have a piece of alder. Notice how much farther apart the grain lines are on this piece of wood. Alder is a much faster growing tree than oak, and so the yearly growth rings are thicker than they are on an oak tree. Alder is not only much faster growing, but it also has a shorter life span than oak, so it also has a different reproductive strategy. Alder is one of the first trees that grows back after a clear cut or fire in the Pacific Northwest. The alder is deciduous, so its leaves form the humus on the forest floor that serves as a mulch bed for the fir trees that are going to grow up next to the alder. The fir trees don't grow quite as fast as the alder, nor do the redwood trees in Northern California forests, but at about the time the alder trees are reaching the end of their life span, the Douglas fir or coastal redwoods are already healthy and have reached above the tops of the alder trees.

Looking at this from a composites designer's perspective, the oak tree has a different set of design needs and requirements than the alder tree, leading to a different combination and fabrication technique using the same string and largely the same glue. This is why I like the tree analog so much when describing composite materials and why I like the semantic tree metaphor as a means to view learning. Growing up in a Pacific Northwest rain forest didn't hurt, either.

Oyster shells—another example of Mother Nature at work—are also a composite material. Thin layers of calcium carbonate (limestone) are laid down and glued together with a protein and polysaccharide (literally many sugars) material (the glue) to form a laminated structure. Pearls are just a grain of sand or a bead of some sort that the oyster surrounds with its layers of calcium carbonate, the binding protein, and polysaccharide glue in many layers. The color of the pearl comes from the other minerals besides calcium in the local waters where the oyster lives.

Animal bones are a composite material also. They are made up in honeycomb-like matrix structures made of collagen, which is a very common protein that makes up the ligaments and most of the cell walls in animals. The honeycomb matrix is filled with small irregular calcium phosphate crystals in a very intricate pattern. This bone is remodeled constantly during the life of the animal depending on the activity of the animal and the stresses induced in the bone. Distance runners are keenly aware of this remodeling of bone—especially in their lower legs, and especially when they get shin splints or even stress fractures of their shin bones. Take it from me, it hurts. But when everything heals back up, as long as you don't over-stress it during the healing process, your shin bone is much stronger because your bone has remodeled due to the induced stress of constant pounding on your lower leg.

There are numerous times that man has copied Mother Nature to make things that go faster, are lighter, stronger, and stiffer, hold more things, save fuel in both cars and airplanes, jump tall buildings with a single bound—oops, got ahead of myself. But you get the point. The history of composites is replete with stories of curious engineers (and others who like to make things) taking a good long look at what Mother Nature has made to come up with solutions to the problem they are trying to solve. And they have worked on that problem until they have created a new material or a completely new way to make something that is useful.

Evidence of the first use of string or fiber to reinforce another substance to make a stronger material comes from archaic humans who used straw mixed with mud to make blocks for building longhouses. These mud bricks were a composite material, but not a completely man-made composite like concrete, since these early humans were just basically using what they could collect from riverbanks where grasses grew in the mud. There is evidence of this in the archeologic records dating back to early humans nearly 12,000 years ago.[2] The people of Jericho used these mud bricks (*mudbrick* or *adobe*, depending on where they are found) to build their houses. The mudbricks were a mixture of loam, sand, and clay as the glue, and straw or long sticks as the string. They were dried in the sun into shapes that could be used for building a hut. By about 3000–2500 years ago, these bricks were fired to make a primitive fired brick that would withstand monsoon flooding. The fired bricks started to be used for the bottom of the building so it would not collapse in the monsoon floods.

Lightweight composites came into use around 3500 BC in Mesopotamia, where wood strips were glued together with a polymer pitch extracted from the bark of trees. The layers of wood strips were placed at different angles to make a form of plywood.[3] Originally, the Mesopotamians did this because of a scarcity of good wood. The lesser quality wood was placed in the middle layers while better wood was placed on the outer layers. This plywood was used primarily as a building material. Actual plywood of today uses much the same idea: thin layers of wood peeled from trees are glued together in line with the grain of the wood—the direction of the strings—then rotated by 90° in each layer until a layered composite panel is made.

In early Egyptian history, starting circa 2200 BC, layers of linen fabric (flax fiber woven into a cloth) or papyrus were soaked in plaster and layered into a mold to make death masks.[4] This is the first evidence of a composite material made by molding a reinforcement in layers (linen or papyrus) and binding it together with a glue (plaster) to make a molded surface with a particular shape. This is basically the same process that is used to make airplane wings, car fenders, wind turbine blades—the list goes on and on.

This concept of using layers of strings (fibers) laid down in different orientations and bound together with glue (resin) to form a solid material that is stronger and stiffer in each direction than either the strings or glue are by themselves is the fundamental basis for all string and glue–type composite materials. You will keep hearing this repeated over the course of reading this book because it is one of those fundamental concepts that needs to be ingrained in any engineer or maker who wants to start to play with composites. The concept to be understood here is that the fiber orientation in any particular layer matters because the layer is strong and stiff only in the direction of the fibers, and it is fairly weak across the fibers. Modern composite designers not only understand this, but it is built into their DNA, so to speak.

The history of modern composites, interestingly enough, goes back to the initial work by Leo Baekeland (Bakelite) after he had become quite wealthy from his invention of a photographic plate that could be developed using water rather than some of the caustic chemicals used in early photography.[5] Baekeland, originally from Ghent, Belgium, was a budding and brilliant chemist from an early age. Growing up the son of a cobbler in Ghent, he graduated with honors from the Ghent Municipal Technical School and was granted a scholarship by the City of Ghent to go to the University of Ghent, where he finished his PhD in chemistry at the age of 21. He and his wife went to New York on a travel scholarship and met Professor Charles Chandler at Columbia University who introduced the young PhD chemist to Richard Anthony of the E. and H.T. Anthony Photography Company. After leaving Anthony Photography and attempting and failing to become a consulting chemist, Baekeland turned back to photography and invented a photographic paper that would allow enlarged photographic images to be taken using artificial light and printed on photographic paper, which he called Velox. Prior to Baekeland's invention, photographs that were printed on paper had to use bright natural light and long exposure times, along with some very nasty chemistry. Since this occurred during the depression, his invention did not gain much traction until he hooked up with Leonardi Jacobi, a venture capitalist, and created the Nepera Chemical Company in Nepera Park, Yonkers, New York. In 1899, Baekeland was invited to meet George Eastman who immediately offered him a million dollars for his Velox photographic paper, which Baelekand accepted. Eastman went on to form Eastman Kodak, which became the largest photographic film manufacturer in the world.

Now that Baekeland was independently wealthy, he had the time and resources to devote to his passions. He began experimenting with phenol (benzene alcohol, a.k.a. carbolic acid) and formaldehyde mixed together and heated to a high temperature in a condensation reaction (we will get to these sorts of reactions later in the book, but for now just follow along with me). Chemists at the time had begun to recognize that many natural resins and fibers were polymers (more about this when we get to glue), and the race was on to create a man-made polymer. Actually, Baekeland was looking for a replacement for shellac since that was a natural material made from the excretions of lac insects, and it was in short supply. He made one called Novolac—literally, new lacquer—that was not a commercial success, so he started experimenting with impregnating wood with phenol and formaldehyde rather than coating the wood with his plastic, and then heating and applying pressure to create a hard, yet moldable material called Bakelite, the first commercial plastic. Bakelite found instant success for use as an electrical insulator that was lighter, nearly as strong, and much less brittle than the fired ceramics that had been used. He filed his first Bakelite patent in 1907, and the plastics industry was born.

Baekeland went on to form the General Bakelite Company and made a transparent plastic initially intended to be laminating varnish, but eventually the markets for both a casting resin and a molding resin outstripped the varnish market. And there was increasing competition from other manufacturers of phenolic resins—specifically Condensite Company of America founded by Jonas Aylsworth, and the Redmanol Chemical Products Company formed by Lawrence Redman. In 1922, after Baekeland successfully litigated his patents, Aylsworth and Redman agreed to join Baekeland to form the Bakelite Corporation. Eventually the company was acquired by Union Carbide, which has since been acquired by Dow Chemical and is still making Bakelite today. In England, the rights to Bakelite were acquired in 2005 by Borden Chemical, now Hexion, which makes phenolic, epoxy, and other casting resins.

The Bakelite story is typical of the glue part of the composites industry, where new resins typically start out in some brilliant chemist's garage or private laboratory. Once they had something that they thought they could sell, the chemists would approach a chemical company with their patent to commercialize it. For instance, epoxies[6] were originally formulated and patented by Paul Schlack of Germany in 1934, during the reign of Hitler. That invention didn't make it to the Allied side of that war, but Pierre Castan of France was working on much the same thing, without knowing about Schlack's work, and in 1940 he patented his own epoxy. In 1943, Castan was working with bisphenol-A, and found (or claimed?) that he could make a suitable resin from this organic backbone. He licensed his work to Ciba, which went on to become one of the three largest epoxy resin manufacturers in the world. In 1946, Sylvan Greenlee, working for the Devoe and Reynolds Company, patented a resin developed from bisphenol-A and epichlorohydrin. Devoe and Reynolds eventually sold to Shell Chemical, which eventually became Hexion, one of the largest epoxy manufacturers in Europe.

Fiberglass or glass-reinforced plastic (GRP)—which is the material comprising pretty much every recreational boat hull as well as all surfboards and many other things—has a history that closely follows two major industries, glass making and organic resin manufacture. In 1888, Edward Drummond Libbey, often referred to as the father of the glass-making industry, formed Libbey Glass Company in Toledo, Ohio. His new company made glass bottles using equipment invented by Michael Joseph Owens. In fact, Libbey's success relied heavily on Owens' inventions, and in 1903, Libbey founded the Owens Bottle Machine Company. Owens Company made its first splash on the world market in an 1893 World's Columbian Exposition in Chicago, where Libbey showed off a dress made entirely from glass fiber that was as thin as silk using an Owens invention. Libbey and Owens had entered into a partnership after Owens invented a machine that could make glass bottles at the rate of 240 per minute, and the resulting company became the Owens-Illinois Glass Company. Libbey went on to form the Libbey Sheet Glass Company which expanded to have five glass manufacturing facilities in the United States, including one in the City of Industry, California. Owens-Illinois is the company that eventually merged with Corning to become what we know today as Owens Corning. Libbey passed away in 1925, but the Libbey Glass Bottle Company and Owens-Illinois lived on.

Back to glass fibers. There wasn't a particularly great demand for glass fiber until 1933 when Games Slayter, who was working at Owens-Illinois at the time, invented the glass wool that all of us use as insulation in our houses.[7] Slayter was originally hired as a consultant to the Owen-Libbey Glass Bottle company and was hired to develop glass blocks for the construction industry. When he was working at the plant, he noticed long glass fibers hanging from the ceiling and thought that they might make a good filtration medium. He proceeded to experiment with these fibers and made the first commercial fiberglass product—the DustStop air filtration system. Prior to Slayter's invention, which went into commercial production in 1936, glass fibers were sold in a *staple*, which was a bunch of relatively short fibers that were individually stretched. The Corning Glass Works got interested in the invention of continuous glass fiber being spun into glass wool (fiberglass), and Owens-Illinois merged with Corning Glass Works to become Owens Corning Fiberglass Company. See, I got you back here. This new invention of continuous glass fiber, or *filament* fiber as it is now known, was first mass produced by Owens Corning when the two companies merged in 1938. This is the glass fiber that we know of today in the composites industry, and Owens Corning is still the major producer of high-performance glass fiber in the market today.

The typical glass fiber that is in use in the composites industry today is called E-glass—or electrical glass—because it has low electrical conductivity. E-glass is an alumino-borosilicate glass that has less than 1 percent alkali oxides, giving it good electrical resistance. It also is very inexpensive to make and has great flexibility, relatively high strength, and relatively high toughness. It is, however, susceptible to chloride ion attack whereas E-CR glass (electrical-chemical resistant) glass is not. Anyone who has owned a larger recreational boat that stays in the water all the time (larger sailboats and power boats) knows that they need to haul out their boats every few years to have the bottom recoated and to look for what are called *osmotic blisters*. These are created by small cracks or defects in the gel coat (we will get to these resins later in the book) that allow seawater to get to the fibers. Sea water, of course, has abundant chloride ions, since salt is sodium chloride. The osmotic blisters are the result of the corrosion of the fibers, which creates acids that eat away at the resin. And, if left for too long, the acids will eat all the way through the hull of the boat.

At about the same time that Owens Corning began mass production of glass fiber, Carlton Ellis of DuPont de Nemours invented a polyester resin. He was awarded a patent for it in 1936. At the time, the Germans were working on the same things as DuPont and they went on to further develop polyester resins. By then the Second World War had broken out and DuPont moved on to other things like gunpowder and explosives. However, World War I British spies went into Germany, stole the secrets of the Germans, and turned them over to American firms—Cyanamid in particular, which eventually became American Cyanamid. In 1942 American Cyanamid produced the first truly commercial polyester resin, which is the direct forerunner of today's polyester resin systems. And as early as 1942 Owens Corning was producing fiberglass/polyester airplane parts for the war effort. In 1937 Ray Greene of Owens Corning built a fiberglass boat, but he thought the resin he used was too brittle and not quite up to the task. But, after a few years of experimentation with different resins, he settled on the American Cyanamid version of polyester and made a daysailer in 1942 using fiberglass. This was the first production of a fiberglass boat, and it spawned the enormous industry we all know today as the recreational boat industry.

Today, nearly everyone in industrialized society owns something made of fiberglass. It is in kitchenware, shower stalls, bathtubs, bath faucets, auto body panels, truck bodies, fishing rods, tennis rackets, bicycle frames, even large storage tanks for liquids as well as liquified gases, and the list goes on. Fiberglass is ubiquitous as insulation in all our newly built buildings. It is even in some bridges that you drive over every day in the form of fiberglass rebar. And if you take an airplane ride on a commercial jet today you are flying in something made at least partially from fiberglass. In fact, the Boeing 787 is 50 percent composite, with a large portion of that being fiberglass.

Carbon fiber[8] has a similar history to glass fiber. Short glass fibers were made for a couple of centuries before Owens' invention produced a continuous filament glass that we know today as fiberglass. Similarly, Sir Joseph Wilson Swan first created short carbon fibers by heating threads in an inert atmosphere to carbonize them. In 1897, Thomas Edison used this technique with cotton threads and thin strips of bamboo to make the first electrically powered light bulb. These "Edison bulbs" were superseded by tungsten filament bulbs in the early 1900s to create the common incandescent bulb of today, and carbon fiber production languished for more than a half a century. Then, in 1958, Roger Bacon, working at the Union Carbide Parma Technical Center in Cleveland, was searching for the triple point of carbon when he discovered that short carbon whiskers or fibers were forming on the negative electrode of his arc furnace when he heated strands of rayon in argon. These fibers were only about 20 percent carbon, and they were not suitable for any of the applications for carbon fiber that are in use today. In addition, Bacon's process was very inefficient and not commercially viable. Then, in the early 1960s, Dr. Akio Shindo of the Agency of Industrial Science and Technology in Japan exposed fibers made from PAN (polyacrylonitrile) to intense heat in an argon atmosphere and created a fiber that was 55 percent carbon and was commercially viable.

PAN[9] itself also has an interesting history. Polyacrylonitrile has the chemical formula $(C_3H_3N)_n$ where repeating groups of acrylonitrile (the C_3H_3N part) are reacted together to form a long chain. Physically it is classed as a *semi-crystalline* organic polymer resin. This is because it forms crystal-like structures when it is dried out of its water-based or aqueous solution. However, it does not melt, and instead will decompose when heated. It was first synthesized in Germany by Hans Fikentscher and Claus Heuck in the Ludwigshafen works of the German chemical conglomerate IG Farben.[10] However, since PAN doesn't melt, and was not soluble in any of the organic solvents they tried, further research with it was halted. Then, in 1931, Herbert Rein, head of polymer fiber chemistry at the Bitterfield plant of IG Farben, discovered that PAN would dissolve in pyridinium benzylchloride. He spun the first fibers made of PAN in 1938 using a mix of different water-based solvents. But then the Second World War intervened and commercial production was halted. The Brits stole the German secrets to making PAN fiber just as they had stolen the German secrets to making polyester resins, and DuPont first mass-produced PAN fiber in 1946. The DuPont patent for their new product, called Orion, was filed exactly seven days after a nearly identical German patent.

At the same time that Dr. Shindo was experimenting with PAN, Curry Ford and Charles Mitchell were working at Union Carbide's Parma Technical Center. In 1959 these two patented a process for making fibers by carbonizing rayon fibers at temperatures as high as 3000°C. This produced the strongest commercially viable carbon fibers to date and eventually begat the advanced composites industry in the United States in 1963. At the time, the researchers at Union Carbide didn't know about the work going on in Dr. Shindo's lab, partly out of hubris, partly because the U.S. chemists had not been able to produce a high-elastic modulus, high tensile strength fiber from PAN, and partly because the Japanese were keeping quiet about their work. The U.S. military got involved and quite a bit of the work on creating higher elastic modulus carbon fiber at Union Carbide was classified because of its potential use in nose cones of re-entry vehicles. This was the Cold War, after all.

While the U.S. fiber producers were infatuated with rayon and what they could make with it, chemists and fiber scientists in both the UK and Japan were working on PAN fibers. This was largely because, at the time, they had pure PAN to work with whereas U.S. companies that made PAN fiber intentionally doped the PAN with other impurities (creating copolymers of PAN) to modify the fiber properties of the fiber for the textile industry. This, however, rendered American PAN fibers nearly useless for making high-strength, high-elastic modulus carbon fiber, mostly because of all the impurities added to and co-polymerized with the PAN fiber to make sweaters and acrylic socks in nice bright colors and to make the fabrics soft and wearable. Eventually the Japanese beat the Brits at their own game, becoming the most successful PAN-based carbon fiber producers in the world. The United States finally noticed, and in 1970, Union Carbide signed a licensing agreement with Japan's Toray Industries and the United States was finally back in the game of making high quality, high strength PAN-based carbon fiber. Toray and Toray-licensed PAN-based carbon fiber accounts for most of the carbon fiber in composites to this day.

Others were working on making carbon fiber from precursors other than rayon and PAN. Leonard Singer came to the Parma Technical Center in Cleveland[11] in the mid–1950s and began experimenting with electron paramagnetic resonance, which is a fancy name for watching atoms move around in a strong, fluctuating magnetic field to figure out how they are geometrically arranged. He was studying the underlying mechanisms of carburization, using various petroleum and coal-based materials. Heating these materials produces a pitch, commonly known as coal tar, that has a number of different cyclic carbon compounds all mixed together. Pitch is an important precursor to many graphite and other carbon-based industrial products, so these were especially prominent in Union Carbide's research efforts.

During the original heyday of carbon fiber work when Singer and Cherry were working in Cleveland, a couple of Australian scientists found out that if they used what is called a melt-spin process, they could orient the 6-carbon aromatic rings along the direction of the fibers they were pulling out of the black, tarry pitch. In its natural form, pitch is what is called *isotropic,* meaning that it has the same properties in all directions. What Singer saw in this was what he called a *liquid crystal* state wherein the liquid pitch was ordered preferentially in one direction. With all the other fiber research going on all around him, he inevitably came across the idea that he could pull a fiber out of this liquid crystal—what he called a *mesophase.* Singer and his assistant Allen Cherry tried elongating this mesophase by essentially designing a taffy-pulling machine that aligned the mesophase liquid crystals into an elongated fiber, which when heated, created a very high-modulus and highly ordered carbon fiber.[12] This fiber was the first iteration of what is called *graphitized* carbon fibers because the fiber has a regular structure of 6-carbon rings (benzene rings—remember this, it's important later in the book)—a graphitic structure. This new fiber had amazing strength (in the laboratory), ultra-high elastic modulus, and surprisingly high thermal conductivity and electrical conductivity along the fiber axis. In mass production, these fibers are called high-modulus fibers, and the higher thermal conductivity fibers are used in aircraft brakes where their stiffness and rapid heat removal from the braking surface make them the material of choice. Pitch-based carbon fibers are also used in satellites where thermal conductivity is extremely important because the satellite components cannot be air cooled. They are, however, brittle and somewhat more expensive than PAN-based carbon fiber, so still today, structural applications of carbon fiber are dominated by PAN fiber.

I need to define two more terms for you so that you understand them a little better:

- **Thermal and electrical conductivity** are measures of the ability of a material to conduct either heat or electricity with little resistance. Copper has high thermal and electrical conductivity so it is used for the electrical wires in your house and also on the bottom of your cookware so that the soup pot heats evenly when you put it on the stove. We will talk a little more about this in the next chapter when I talk about the properties of carbon.

That's the basic history of composite materials that we know of today—at least an introduction to their history. Now we need to start learning about materials, as in fibers and resins. Before I introduce materials, first we need to understand where these materials come from. The answer to that comes in the next chapter—it's all about the periodic table of the elements and what that table means to the practicing composites engineer.

2

Composites and the Periodic Table of the Elements

I bet you didn't know this, but carbon was the third element produced in the Big Bang, after hydrogen and helium. And carbon is the fourth most abundant element in the universe after hydrogen, helium, and oxygen. And you may not have known this either, but carbon is the most important element in all advanced composites, whether it's the fiber or the resin, or both.

But first a little introduction to the periodic table of the elements (or review for those of you who took chemistry not too long ago to still remember it). This is intended for someone who wants to understand the nature, properties, chemistry, mechanical behavior, and *why bother* of composites. It is the dictionary that all good composites engineers use to develop the wonderful things that are developed using advanced composite materials. This dictionary is the main part of the trunk of the semantic tree of knowledge of composites—understanding the periodic table and the implications of how each of these atoms interacts with the other atoms that make up the strings and glues of composites. Let's get busy learning the dictionary.

A Russian chemical genius named Dmitri Mendeleev organized all the known elements into what we now know as the periodic table and published it in 1869.[1] At the time there was tremendous competition in physics and chemistry to discover new "elements" and organize them into recognizable groups based on any number of physical or chemical properties. And at the time there was even controversy over what could rightly be called an element (see Disappearing Spoon[2]).

Note here that the lighter weight elements—lower atomic number and therefore lower atomic weight—are all at the top of Mendeleev's table. Interestingly, these elements, with the singular exception of carbon, were all discovered in modern times (AD vs. BC), in stark contrast to heavier metals. Copper is thought to be the first element discovered around 9000

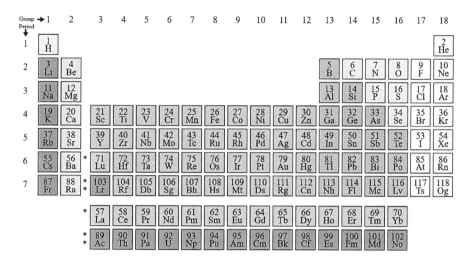

Periodic table of the elements (produced by Offnfopt, Wikipedia Creative Commons License).

BC by Bronze Age humans in the Middle East.[3] Carbon itself was discovered by the ancient Egyptians and Sumerians because they used it in the form of charcoal for the reduction of copper, zinc, and tin ores. Carbon was not, however, chemically identified as an element until the mid–18th century, and was listed finally as an element by Antoine Lavoisier[4] in 1789 during the heyday of element discovery. From the late 1700s and early to mid–1800s there was intense competition among chemists and physicists to put their names on their discoveries.

The elements most important to composites—all of which are at the top of Mendeleev's table—were discovered in the period between 1766 and 1772. Hydrogen was discovered in 1766, oxygen in 1771, and nitrogen in 1772.[5] By "discovered" I mean that they were isolated, purified, and discovered to be separate from the other elements per Mendeleev's initial definition of an element. There is one more element, discovered in 1823, that is important to composites and is the second most abundant element in the earth's crust: silicon.[6] Silicon is right below carbon and is in the next group of elements on Mendeleev's table, and it has a chemistry that is very similar to carbon because it also sits in the middle of its group. Silicon dioxide, or silica, is one of the most abundant molecules on Earth—quartz makes up 10 percent of the earth's crust—and it is the major component of rock and beach sand. Silicon forms compounds with very complex structures just like carbon but is heavier, so it makes a very good element for making building materials. Think concrete, stone, mortar, glass. It is glass that I'm thinking of here, which is important to composites in the form of glass fibers.

While there are some other elements that are also important to composites, these are the basic ones. We all know that modern composite materials were originally developed with the goal of making a material that could replace metals with something lighter but just as strong and stiff as the metal it replaces. What is less understood is that this is why all the important elements are at the top of Mendeleev's table. His table is organized by atomic number or number of protons in the nucleus of a single atom of the element. And because of the laws of physics, the atomic weight (protons + neutrons) closely follows the atomic number. And atomic weight is a fair estimator of the density of a solid pure element. For example, a cubic inch of carbon weighs less than a cubic inch of silicon, which weighs less than a cubic inch of iron, and so on with lead being quite heavy.

Protons, Neutrons, Nucleus of an Atom, Electrons, Atomic Number, Electron Orbitals, Valence Electrons—What Are These Things?

In 1909, Ernest Rutherford's[7] student, Ernest Marsden, got a very strange result from an experiment he was doing in Rutherford's lab. He found that some particles coming off a radioactive source had bounced off a gold foil (this is called backscatter in these experiments), rather than just slowing down and diffracting on their way through the foil. The model for the atom that physicists believed in at the time was called the *plum pudding model,* where electrons were thought to be swimming around in a thick, positively charged viscous mass of stuff. Marsden re-ran the experiment several times and got the same result—about one in a few thousand of these particles (alpha particles) actually bounced backward off something solid rather than just having their paths bent on the way past the thin foil. This experiment is credited with the discovery that the bulk of the mass of an atom resides in a clump in the center of the atom, the nucleus, which is made up of protons (positively charged particles) and neutrons that don't have any electrical charge at all. Rutherford ran some simple calculations and found that the nucleus of the atom was on the order of 1/100,000th the size of the atom. Later, in 1911, another one of his graduate students, Niels Bohr,[8] a quantum physicist by training, started thinking about the role of the electron in atomic physics and also in chemistry. He was trying to rectify the size of the nucleus versus the size of the atom, and found that electrons had to be in discrete energy states for atoms to stay intact rather than falling into the center and collapsing the atom. He eventually proposed what is called the Bohr model, in which electrons were in orbitals circulating around the nucleus but were only permitted to have discrete values of an energy state called *angular momentum*. What this meant is that they were all in discrete energy states or energy levels based on which orbital they were in. When Bohr applied his model to the elements in the first two rows of Mendeleev's table, he explained the basic chemistry of these elements as we know them. The first—or closest—orbital is only permitted two electrons in Bohr's model. That begets hydrogen and helium. We all know that hydrogen is a very reactive gas. When it reacts with oxygen, we not only get water to drink, but we also can fly people to the moon and back. And we also know that helium doesn't chemically react with anything—it is the lightest of the noble gases. A cartoon of Bohr's model that covers most of the first three rows of Mendeleev's table is shown below. This is phosphorus because it has an atomic number (number of protons) of 15.

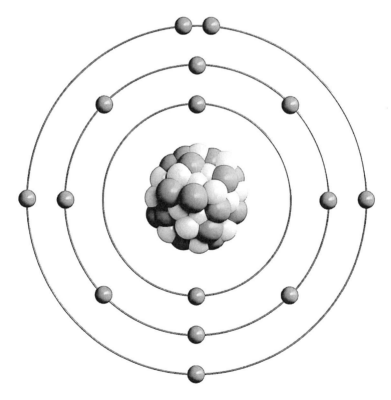

15 Protons   16 Neutrons   15 Electrons

Atomic nucleus and first three electron orbitals (oorka/iStock).

If you look at this picture, you see that there are two electrons in the first of Bohr's orbitals outside of the nucleus (the red ring with the two electrons). Orbitals are often called *electron shells* so if you see them called that in other books on chemistry, you know what they mean. This first row is the row of hydrogen (1 electron) and helium (2 electrons). Bohr said that helium's two electrons are in the first of the stable energy states, which means that they are quite comfortable staying in that state forever and are not looking to either grab an electron or give one up. The next orbital out contains 8 electrons in its fully populated or stable form. This is the group from lithium (3 electrons) to neon (10 electrons) and includes beryllium (4), boron (5) (actually from borax, not Niels Bohr—his is Bohrium–#107), carbon (6—our friend), nitrogen (7 electrons and 80 percent of what you breathe), oxygen (8 electrons and the part of air that keeps you alive), fluorine (9), and finally the noble gas, neon, which filled all the early neon signs. Both helium and neon are called noble gases because their electron orbitals are full and they don't have any spare electrons to use to react with anything else, nor are they looking for electrons to complete their electron orbitals. They are quite comfortable being lone atoms floating about and will completely ignore other atoms.

What does all this have to do with composites, you ask. Well, the elements that make up the majority of the matter that composites are made of are primarily in the middle of this first row of 8. But first, I need to cover a little bit about chemistry and the difference between ionic and covalent bonding, so that you understand why the most important element in composites comes from the middle of this row. I am going to use a simple example of an ionic bond so that you understand how this works. Think of common table salt—sodium chloride. Sodium has 11 electrons and is on the left side of the third row in Mendeleev's table. In its third orbital it has one lone electron floating around looking for something to react with or to be attracted to. Chlorine has 17 electrons and sits over almost to the right end of Mendeleev's table, so it is missing one electron in its third orbital and is desperately seeking to fill up that third orbital. When you put a sodium atom and a chlorine atom together, the lonely electron on the sodium atom finds a very welcome home in the third orbital of the chlorine atom.

We all know that salt dissolves very readily in water because all the oceans are salty, and we use a pinch of salt in pretty much everything we make—including your breakfast oatmeal. That is because water is what is known as a *polar* solvent, which means that it has a side that is positively charged and a side that is negatively charged. The figure below shows what this means.

You can see in this picture that the oxygen side of the water molecule is slightly negatively charged and the hydrogen side is slightly positively

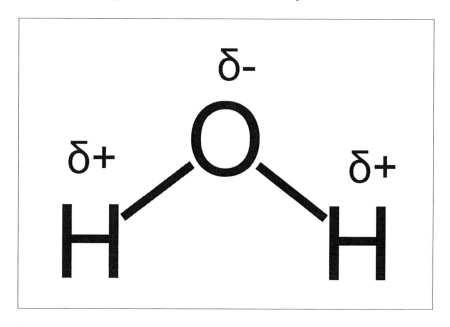

Water molecule—negative side is oxygen; positive side is hydrogen (drawing by the author).

charged. This is because there are many more protons in the nucleus of the oxygen atom than the single lone protons of the hydrogen atoms, so the attraction of the oxygen nucleus to the hydrogens' electrons is much higher than the attraction of the hydrogens' lone protons to the hydrogens' electrons. And when a water molecule (actually several water molecules) comes in contact with a sodium chloride molecule, the sodium atom is attracted to the positive or hydrogen side of the water molecule and it attracts one of the lone protons. In chemistry we talk about charges on ions, so for chemists to understand this, they talk about the sodium ion as positively charged (they call it Na^+), when it picks up one of these protons or shares its electron with the water molecule. The chlorine atom then needs to look for an electron to fill its third shell or orbital, so it is attracted to the oxygen side of water and it grabs one of the electrons from that side of the molecule and becomes what chemists call a negatively charged ion, or Cl^-. While these ions can't live outside of a water solution, this is at least an easy way to understand what an ionic bond is. Ionic compounds dissolve readily in water.

The atoms at either end of the 2nd and 3rd row of Mendeleev's table, and even the 4th, 5th, 6th, and 7th rows, with the exception of all the noble gases, all form ionic bonds. But what do the elements in the middle of each of these rows do when they form compounds? They form what are called

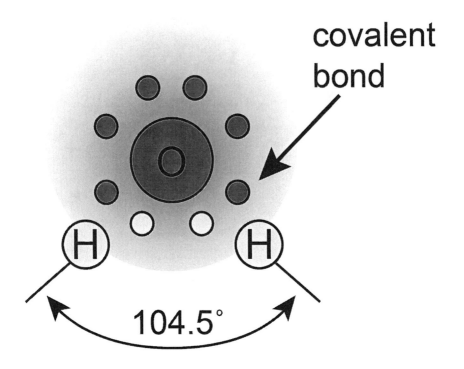

covalent
bond

104.5°

Water showing the electrons it shares (covalent, dark shaded circles) with hydrogen (H, light shaded circles) (*Chem1* online textbook, reproduced by permission of author Stephen Lower).

covalent bonds. This means that rather than one atom giving up an electron (or two) to another atom to form a chemical compound, these elements share their electrons when they form compounds. Oxygen is in the middle part of the second row of Mendeleev's table and it forms covalent bonds—even with hydrogen to form water. Here's another picture showing the covalent bonds that make water.

Atoms that make covalent bonds form very strong chemical bonds that are difficult to break. Water is notoriously difficult to break up into hydrogen and oxygen, which is why water is one of the more abundant molecules in the universe, especially in areas where stars and planets are forming. Hydrogen is of course the most abundant element in the universe, and oxygen is right up there with carbon, so when clouds of gas and dust first start to collect to form stars and planets, the hydrogen and oxygen find each other easily and form water. This is also why astronomers and cosmologists use the signature of water to search for new worlds.

Carbon is the king of covalent bonders because it shares 4 of its 8 electrons in its second or outer orbital—it's smack dab in the middle of that first row of 8. And that is why carbon is the structural element of all life as we know it. The elements that are the most prevalent in organic composites are carbon, oxygen, nitrogen, silicon, and hydrogen. (Where would we be without hydrogen, the stuff of stars?) Carbon, nitrogen, oxygen, and silicon prefer to form covalent bonds. This covalent bonding is what makes composites so strong and stiff.

Another aspect of covalent bonds is that compounds made of these elements can have very complex structures and shapes, as well as complex chemistry. As any biology major can tell you (I was once), organic chemistry is a back-breaker—a weed out class, whatever you want to call it. It was tough. That is because organic chemistry is really the chemistry of carbon and its interaction with primarily hydrogen, oxygen, and nitrogen (and sulfur, of course, but sulfur isn't that important to composites). It is the shape of the carbon-based molecules (proteins, fats, sugars, etc.) in our bodies that gives us the ability to do all the things that we do as humans. And those shapes are largely due to the carbon atom's 4 valence electrons and the different and myriad ways that carbon interacts with other elements to form compounds.

One of these shapes that is critical to not only the resins but also carbon fibers is the ring structure made up of 6 carbon atoms that is called benzene. The figure below shows three different representations of this relatively simple looking compound.

While this looks like a simple enough structure to understand, once this ring gets bonded to something else the shape and chemistry become very complex. This structure is in fact the basis for all resins that are used in the composites business. It is also the shape that makes carbon fiber as strong as it is.

Three representations of benzene (drawing by the author).

Carbon

Let's return to carbon and its chemistry for a little bit. Here is where you find out why I make such a big deal out of the chemistry, atomic structure, and molecular shape of carbon and its compounds. In other words, to know carbon is to know composites and to know composites you have to befriend carbon in a big way.

Carbon has many different physical forms (called allotropes) that are all distinctly different from each other. We all know that the lead in your pencil is pure carbon, and that rock on your wife's or mother's wedding ring is also pure carbon. So why is one black and soft enough to write your name with, whereas the other is the hardest naturally occurring mineral— hard enough to cut steel? That is the magic of carbon. Graphite (literally, rock that writes), commonly used as pencil lead, is black and soft, and when it occurs naturally it comes in a flaky form of layers that slip along one another. It is also a very good electrical conductor, so good that carbon arc furnaces use graphite electrodes to melt iron and steel. And graphite does not melt, nor is it very easy to light on fire. It can withstand temperatures in excess of 3000 degrees when other materials are long gone. The crystal structure of graphite is layers of these 6-carbon rings loosely bonded together. Diamond, on the other hand, is a poor conductor of electricity—it's actually a semiconductor—but it has very high thermal conductivity whereas graphite, if the rings that make it up are in the correct orientation, is a very poor transmitter of heat. A comparison of the two different structures of carbon is shown below to give you an idea of why each has the properties that it does.

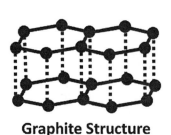

Graphite Structure

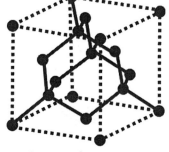

Diamond Structure

Graphite versus diamond showing how graphite forms in layers that can slip along one another, whereas diamond is difficult to break (produced by Diepizza, Wikipedia, Creative Commons Share Alike 4.0 International License).

Carbon, because it can form these complex structures, also forms a wider variety of compounds than does any other element. We are up to almost ten million[9] known compounds of carbon and new ones are discovered or synthesized all the time. And that number is a very small fraction of what is theoretically possible. It's no wonder that carbon is commonly referred to as the king of all elements, or the magic element.

But let's get back to the benzene ring. The most prevalent compound that is a building block of life and of composites is phenol—benzene alcohol. For those who have not had or have forgotten their organic chemistry, an alcohol is a hydrocarbon with an O-H group attached where one of the hydrogens typically sits. Ethanol (vodka, gin, bourbon, etc.—gives you a hangover in the morning if you drink too much of it) is a simple two-carbon chain with an O-H group, like so:

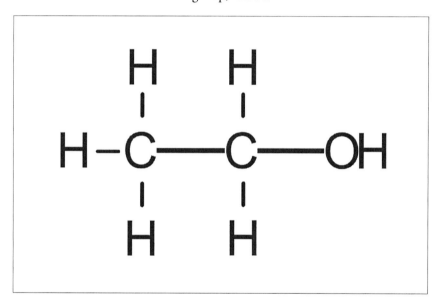

Ethanol chemical formula and structure (drawing by the author).

It is the O-H group (water without a hydrogen) that makes this compound a liquid at room temperature because without this group this is ethane, one of the components of natural gas when it comes out of the well and before it is processed into mostly methane (one carbon atom with 4 hydrogen atoms).

Phenol is the alcohol that benzene makes, and it looks like the following:

This molecule is the backbone of nearly all composite resins, and most industrial plastics. It is also common in most of the plant and animal

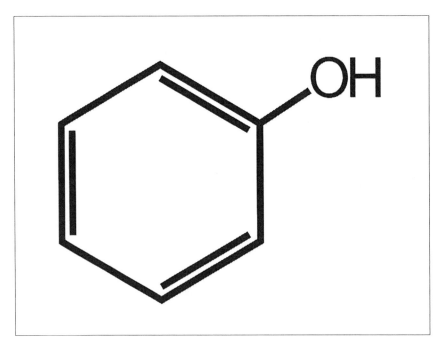

Structure of phenol (drawing by the author).

kingdom. It is the polyphenols in cocoa that make chocolate so good for you—the dark kind, not milk chocolate. Organic compounds that have phenolic groups attached to them are called *aromatic hydrocarbons*, and they live up to that name. Most of them have an aroma, from pleasant to not so pleasant, and a taste, again pleasant to not so pleasant, so they are what give a lot of our foods and spices their flavors. They are what make roses smell so sweet and a skunk smell so bad—sulfur has something to do with the skunk as well, but you get the drift.

Phenol also forms chains (polymers) very easily. Each benzene group is connected to the next benzene group through an oxygen atom. These are called phenolics, which are the first composite resins ever developed. Remember the history of plastics earlier in this book, and Leo Baekeland—Bakelite? Again, more about this when we get into the chemistry of the glue later in the book, but I wanted you to know where this all came from and its connection to the periodic table of the elements.

What about carbon fiber? Well, the ring structure shows up again in all the structural carbon fibers that are made. Graphite comes in several forms, from very regular (graphene) to very disordered, always maintaining that ring structure. This is what is called a *graphitic* structure in materials and chemistry because this repeating series of rings, whether they be in layers or

Turbostratic structure

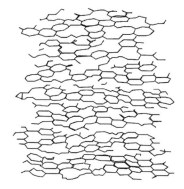

Graphite structure

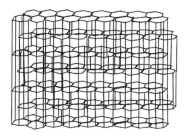

Turbostratic vs. graphitic structure of carbon fiber (14) (produced by Techiescientist.com, Creative Commons License).

all tied together, is a very common pattern in nature. The cellulose in trees and the stems of plants has a ring structure. Even a lot of the collagen in your bones and skin has ring structures. And—famously—DNA is made up of purines and pyrimidines, which are amino acids with ring structures.

These ring structures show up in spades in carbon fibers. There are two basic types of carbon fiber—fibers made from PAN (polyacrylonitrile) which have a semi-graphitic structure that is also somewhat disordered. This structure is called *turbostratic* as compared to the regular crystal structure of carbon, which is called a graphitic structure. The figure above shows the difference between these two structures.

The graphitic structure shown above is stiff and very strong, but is fairly brittle because it can break at one of the planes between the rings. The turbostratic structure isn't as stiff or strong, but it is not brittle and therefore makes a very good structural material that can bend easily and spring back to its unbent shape. The Boeing 787 wing is made of this stuff, and when they tested their first fully test-ready wing, they grabbed the end of the wing and bent it up 25 feet, which was about one and a half times the expected extreme maximum bend that the wing would suffer in the worst in-flight conditions. The test was quite a sight to see, and this is due mostly to the turbostratic structure of the carbon fiber that makes up the 787 wing.

Silicon

Silicon sits right below carbon in Mendeleev's table which means that it also has 4 valence electrons and forms covalent bonds. Silicon dioxide (SiO_2), or silica, is the most common mineral in the earth's crust—it makes up nearly 60 percent of the mass of the crust. In fact, when the earth formed and started rotating, the heavy elements—mostly iron—sank to the center, and the lighter elements—silicon, oxygen, aluminum, etc.—rose to the surface and solidified to make the crust. Silicon, which is the second most abundant element in the crust, is the most important one to the semiconductor industry. Most computer and memory chips are made from silicon because of its abundance as well as its electrical properties. It is what is called a semiconductor, which means that it can be made to conduct electricity by the application of an electric charge, so it makes a very efficient switch. That is really all a transistor is: a bunch of switches etched into a silicon wafer.

Silica, silicon dioxide or SiO_2, again, is the most prevalent compound in the earth's crust and is not a crystal like a lot of other metal oxides. What this means is that it does not have a specific melting temperature. Instead, it softens as it is heated until it becomes a fairly viscous liquid that can be manipulated into a shape. This softening happens at about 1200°C—a temperature that we perceive as very hot here on the earth's surface, but closer to the core of the earth it is about normal. Once it softens, the silica molecules move about freely—thereby making what is called a glass. Silica is, in fact, the major component of all glasses today because of this high temperature softening. Since it is transparent, we use it for windows, bottles, and anything we can make that is transparent. Your cell phone has a high silica content front cover that is electrically conductive, which is why you can touch the screen and make a phone call.

But what does this have to do with composites? Well, since silica (silicon dioxide) is the major component of all glasses, glass fiber is largely made from this stuff. When Russell Games Slayter originally made his invention of glass fiber in 1933 he was working with a soda-lime glass (silicon dioxide with sodium carbonate and calcium oxide) that is typically used to make glass bottles and windows. The glass used for fiberglass has evolved over time to an alumino-borosilicate glass, which is a stronger and tougher cousin of the soda-lime glass that was in use when Slayter made his first long continuous fibers at the Owens-Illinois Glass Company.[10] This higher strength alumino-borosilicate glass fiber is now called E-glass, and it is mostly silica with a mixture of boron, sodium, and aluminum oxides. A typical E-glass has the percentages of these shown in the table below.

Percentages of Silicon Dioxide, Boron Oxide, Sodium Oxide, and Aluminum Oxide in Typical E-glass

Material (Oxide)	Percentage
SiO_2	80.6
B_2O_3	13.0

Material (Oxide)	Percentage
Na_2O	4.0
Al_2O_3	2.3

Note again where all these elements are on the periodic table. Aluminum, boron, and silicon are all grouped right together with carbon in the middle of the top two 8 element rows, so along with oxygen, all of these make covalent bonds which are very strong, giving all silica-based glasses their high strength.

Remember I noted that silicon has 4 electrons that it can share (valence electrons) like carbon, so compounds based on silicon can have very complex shapes. In fact, the crystal structure of pure silicon takes on a shape much like the shape of a diamond crystal, which is a crystal of pure carbon. This structure in crystallography is called a body-centered cubic structure because of the way the atoms are connected. Diamond also has a body-centered cubic crystal structure. The figure below shows this.

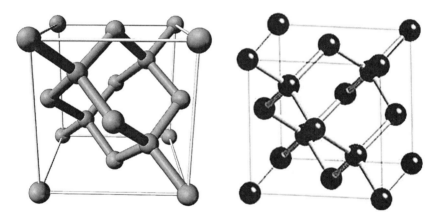

Comparison of the shape of a single crystal of pure silicon (left) to that of a single crystal of diamond (right) (Silicon produced by Ben Mills, Creative Commons License; diamond produced by Kevin Yaeger, Creative Commons License).

If these two pictures look eerily similar it is because they are. This is what you get when you have 4 valence electrons, you are high in the periodic table (low atomic number and low atomic weight) and you can form chemical compounds with complex shapes. A typical borosilicate glass with some sodium ion modification to make the glass more workable has a complex structure that is reminiscent of the turbostratic structure

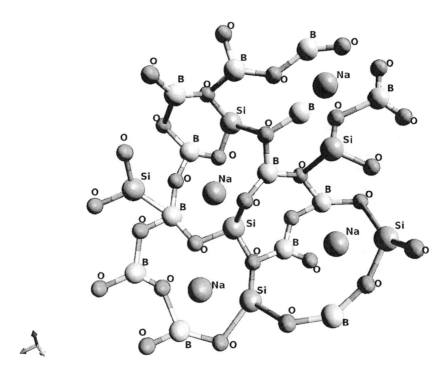

Schematic representation of a sodium-modified borosilicate glass (Dores et al., "Effect of Thermal Treatment on Morphological Properties of Borosilicates Glass Doped with Silver," Encontro de Física Aplicada, Figure 1. *Blucher Proceedings* X, São Paulo, September 2019. Creative Commons License, Attribution 4.0. International).

of carbon fiber. These glasses look like they are a seemingly random collection of atoms, but in fact have a fairly regular structure if you look at a larger grouping of the elements. The figure above is a schematic representation of such a glass. Note the regular occurrence of silicon and boron atoms with oxygen as the glue that binds it all together. Also note the sodium atoms that are sort of dispersed within the matrix of silicon, boron, and oxygen. These sodium atoms make the glass more workable and make the glass have a somewhat lower *glass transition temperature*, or the temperature at which the glass will soften so you can work with it to form it into complex shapes.[11]

This structure, since it is "glassy," is relatively easy to draw into a fiber. It is also made primarily of silicon, which is abundant and very inexpensive. Since glass fibers are easy to make and are really just an extension of the glass bottle and windowpane manufacturing process, they are the least expensive and therefore most prevalent of the composite reinforcing fiber types. Nearly all recreational boats are made of fiberglass nowadays. This is also true of lots of the body panels and other shaped components of cars and trucks that need to be as light and strong as possible while being about the same price as sheet metal.

Oxygen, Nitrogen, and Hydrogen—Elements That Make Up the Glue

Hydrogen is the most abundant element in the universe, followed by helium and then oxygen. And since hydrogen and oxygen are so chemically active, compounds of them outnumber all other chemical compounds on earth. Oxygen is, in fact, the most abundant element in the earth's crust, since the crust is made up of rocks that are themselves made up of oxides of elements heavier than oxygen. If you remember, silicon dioxide is the most abundant mineral (rock) in the earth's crust, and it has two oxygen atoms. What about nitrogen? Well it makes up 78 percent of the earth's atmosphere and is one of the elements found in amines which are typical hardeners for epoxy resins.

Why should you care? Well, all resins—the glue—used in the composites industry are made up of a mix of carbon, hydrogen, and oxygen, with a couple of other elements (chlorine in epoxies,[12] nitrogen as an amine hardener) that modify the properties of the resins to make them suitable for creating a cured composite part. Or, in the case of nitrogen, they are a basic element in the hardeners used to cure the resins to make them strong and stiff. We learned that carbon forms these 6 carbon rings—benzene rings—and when an OH group is attached, it becomes phenol, which is the backbone of most resins used in the composites industry, as well as coatings, paints, most plastics, etc.

We have also already learned about carbon and carbon-hydrogen compounds, or hydrocarbons. These are what we pump out of the ground and burn in our cars, trucks, and airplanes. Oxygen, on the other hand, we have not covered yet—except for mentioning oxygen's role in making alcohols or hydrocarbons with an OH group replacing at least one hydrogen atom. This oxidation reaction is the basis for most of the good stuff that happens in organic chemistry. It is also what enables hydrocarbon-based organic compounds to have such a broad range of properties. Oxidation is, in fact, the very basis for all life on this planet.

One very important molecule in composites is called bisphenol-A.[13] We all know that bisphenol-A, or BPA as it is commonly known, has earned a bad reputation as of late and that the jury is still out about how bad it is. But that's not the subject of this book, so we will leave it at that for

Chemical structure of bisphenol-A (drawing by the author).

now. Bisphenol-A is a very important compound used in the production of quite a few plastics and most composite resins. The structure of BPA should look familiar by now: two phenols linked together in the middle of what appears to be a very short hydrocarbon chain. The figure above shows this structure.

The oxygen atoms at each end of this molecule are what make it so useful in resins and plastics. These oxygen atoms are the sites where polymerization happens. All plastics are polymers of some sort, which means that what is called a monomer (like BPA above) is attached to another monomer, sometimes the same one, and on and on until all the monomers are attached to each other to make up the plastic polymer. This is how composite resins work. The monomers are brought together with a chemical hardener (remember the amine hardeners for epoxies?) and sometimes brought up to a cure temperature, where the reactions happen and all the monomers form up in long chains and harden the resin. This process isn't really all that simple, and a lot of research has been done and is ongoing to improve the properties of these resins and plastics for various purposes. We will learn more about all this when we go over the glue later in the book, but I wanted to introduce you to the concept now so that when we go into it in detail you will be able to follow along.

Finally, what about hydrogen? Hydrogen is a very interesting atom—one proton and one electron, and that's it! But it has sway over one of the most important forces in nature—the electrostatic force. Or is it that the electrostatic force is the only force that is really important to the hydrogen atom? The distinction here doesn't matter. What matters is that the hydrogen atom is very lightweight and therefore its one electron is always in a semi-free state. The only attraction between the hydrogen's proton and its electron is the electrostatic force. The so-called hydrogen bond is an electrostatic attraction between a hydrogen atom that is already in a covalently bonded state with an atom like oxygen and another electronegative atom that has a lone or free pair of electrons—like oxygen (again), or nitrogen, or carbon for that matter. Hydrogen bonding between these chains of polymers is one of the mechanisms that adds toughness to composite resins, so this type of bond is very important to making strong, lightweight composites.

It is, in fact, the hydrogen bond that gives water its unique properties. H_2O is critical for life on our planet, but its behavior is peculiar. Based on molecular weight alone, it is too light to be liquid at room temperature, or to freeze at 0°C, or boil at 100°C. But when you add the electrostatic attraction of one of the hydrogen atoms' electrons to the oxygen atom in a close neighbor water molecule, that attraction causes water to condense and also freeze at a high temperature. The higher the temperature, the more motion and the greater the kinetic energy of each of the molecules. The hydrogen bond in water overcomes this inherent kinetic energy and condenses into liquid water. Hydrocarbons of the same or higher molecular weight (ethane, propane, butane, etc.) are all gases at room temperature and don't become liquids until their temperature is far below the freezing point of water. In fact, they make good refrigerants because of their low gas-liquid condensation temperature. These gases are what is going around in the pipes in your air conditioner, heat pump, and the refrigerator in your kitchen.

Before we leave the subject of hydrogen, we have to learn about acids and bases. We know that hydrogen is the simplest atom and is made up of a single proton and a single electron, so acid-base chemistry is the chemistry of hydrogen. An acid is a proton donor—it has a tendency to want to donate a hydrogen nucleus (proton) and become electrically negative—especially when it is dissolved in water. Here we go with water again, the polar solvent that will dissolve almost anything and forms solutions of salts, acids, basically everything. Hydrofluoric acid (HF) is probably one of the strongest acids known because the fluorine atom is so small that its electron shells are very close to the nucleus. Remember that fluorine sits right next to neon on the periodic table, so its electron structure is searching for one more electron to complete that second shell of 8 electrons. When it finds a hydrogen atom it snaps on to that electron and could not care less about the proton that comes with the hydrogen atom. That's why it donates the proton so easily. Other very strong acids are hydrochloric (HCl) and sulfuric (H_2SO_4), both of which very easily give up an extra proton because their electron shells are complete, and the only thing that's keeping the proton attached is what is called the *electrostatic force*, the force that pulls positively charged things toward negatively charged things. A good example of this wanting to donate a proton can be shown just by the structure of sulfuric acid.

Those two protons out at the ends of the oxygen atoms might look out of place, and you might that the molecule would be better balanced

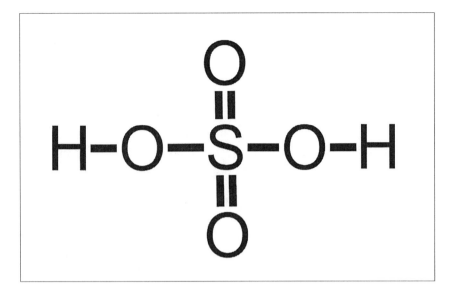

Structure of sulfuric acid with positions of the hydrogen atoms (drawing by the author).

without them, but that is exactly why this is such a strong acid. Those two protons are very nearly free to leave whenever something that attracts them comes along.

One of the strongest bases is sodium hydroxide (NaOH). And again, the chemical structure of sodium hydroxide shows why it wants another hydrogen atom to make it complete.

If only sodium hydroxide could find another proton, it would be water-attached to a sodium. When you dissolve sodium hydroxide in water, there is a lot of bargaining going on about where any stray hydrogen nucleus might become attached to this molecule. In fact, if you were to mix solutions of sodium hydroxide and hydrochloric acid, after all the heat dies down from the violent acid-base reaction chemistry going on, you get salty water—H_2O + NaCl, or table salt.

The reason I go on about acids and bases and the fact that hydrogen is driving all this chemistry is because the acid-base reactions are very important to the synthesis of most all the polymeric resins that are used in composites. Acid-catalyzed reactions and base-catalyzed reactions are the keys to understanding how to stick a bunch of these benzene rings—or in the case of urea-based composites, rings made of nitrogen—together to make a hard but flexible polymer that sticks to fibers when it is liquid and will harden when these catalysts do their job.

So that's an introduction to the periodic table of the elements and its

Na—O—H

Structure of sodium hydroxide (drawing by the author).

importance to understanding composite materials and the properties and chemistry of both the string and the glue. Let's move on to the string and the glue to see what those are made of, how they are made, and how man has been able to harness the laws of physics and the properties of these elements to create fascinating and inherently very useful artificial materials that we all use daily.

3

Composite Fibers—the String

I'm going to start with the string in composites which is the first major branch of our semantic tree of knowledge of composites. I have been leading you to this point for some time, and we're here now. I have already touched on the two most important fibers—glass fiber and carbon fiber—and a bit of their history, manufacture, and structure. Now is the time to delve deeper into each fiber and fiber type, how they are made today, what the different types are used for and why, how to stick resins to them, and how they react when you cure them into a solid. There are also other important fibers that we need to learn about, like Kevlar®, Spectra™, Vectran™, etc. Let's get at it.

Carbon Fiber

We are going to start with carbon fiber because it is the string that is most often used in what are typically called advanced composites, and also because once you understand carbon fiber, all the rest of the fibers come easily. We talked a bit about carbon fiber in the chapter on the periodic table, so let's review that briefly and then expand on it until we have a complete picture of carbon fiber. We need to know how it's made, what its structure looks like and why it has the mechanical, chemical, physical, electrical, etc., properties that it has. We also need to understand why carbon fiber is as wonderfully useful as it is.

When I first introduced carbon to you, I went on at some length about the 6-carbon ring—benzene ring—that forms so many interesting and useful organic compounds both in nature as well as in our industrial society. I also detailed the strength of the covalent bonds in this ring and why it is ubiquitous in nature. I noted how this ring shows up in spades in carbon fiber, and is actually the backbone of all carbon fiber types. It is also the structure of graphite—as in common pencil lead. To begin with the study of the different types of carbon fiber, first we have to understand graphite and why the two different basic forms of carbon fiber are called graphitic and turbostratic. And as always, remember the benzene ring—it is the most important structure of carbon in composites, and the one that is most prevalent in nature.

Graphite

What is graphite and why is it so important? Well, it is the most common form of carbon in and on the earth, coal included. The 6-carbon ring structure in coal, for example, occurs because coal comes from plants that lived millions of years ago—plants that are mostly cellulose and lignin, both of which are made up primarily of 6-carbon ring structures. Graphite is a series of flat layers of these 6-carbon rings all stuck together in sheets. The bonds within the sheets are very strong because they are covalent bonds. Within each ring there are three double bonds and three single bonds. And this is something of a misnomer because the electrons that form these bonds are constantly bouncing around the rings in a dance called *resonance*. This is why this hexagonal ring structure of graphite is commonly shown as a hexagon with a circle in the middle of it. In other words, the double bonds are constantly shifting around the ring—or as some atomic physicists call it, shifting between two resonant structures. People who want to show the structure of graphite (or benzene, for that matter) don't bother with the two resonant structures and just use the hexagon with a circle in the middle to represent it.

When you put a whole bunch of these 6-carbon rings together they form a flat sheet like the structure shown in the chapter on the periodic table. Each sheet formed this way is called *graphene*. Note the -ene in the name of the sheet. To an organic chemist the -ene means that the compound has carbons with double bonds. Then when you take these sheets and pile them on top of one another, you get a piece of graphite—that is what is mined in places where there are deposits of this stuff. The -ite comes from graphite being a naturally occurring mineral, like magnetite,

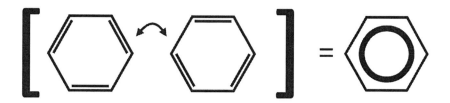

Resonance in the graphitic or benzene structure (drawing by the author).

dolomite, etc., meaning that it is a rock (remember graphite—rock that writes). The bonds within the layers—the covalent, resonant, 6-carbon ring bonds—are very strong and stiff. They are what give the graphitic form of carbon fiber its strength and stiffness. In any case, the structure of graphite looks something like the following figure.

The dotted lines in this figure represent the weaker, van der Waals bonds between the layers, which are what make graphite flaky. What exactly is a van der Waals bond? It is more a force than an actual chemical bond. Van der Waals bonds are very weird. They are either attractive, as in holding sheets of graphene together, or they are repulsive, as in shoving atoms apart, depending on how close or far away the atoms are that are interacting and feeling this force. This happens within a very narrow band of distances between the atoms. When the two atoms are more than 0.6 nm apart—that's nanometers or millionths of a meter—the van der Waals

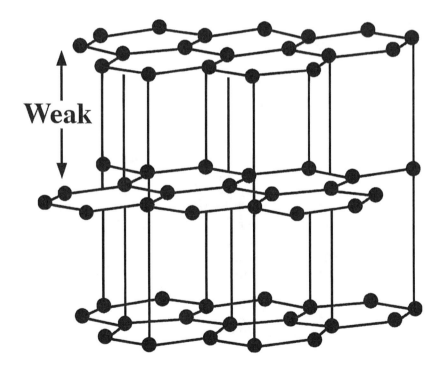

Weak

Graphic representation of the atomic structure of graphite (adapted from Rijesh et al., "Determination of Friction Factor by Ring Compression Test for Al-5Zn-1Mg using Graphite and MoS2 Lubricants," Figure 7, *Thammasat International Journal of Science and Technology*, vol. 17, 2012, pp. 13–19. Open access).

forces disappear. When the two atoms are between 0.4 and 0.6 nanometers apart, the van der Waals forces attract the two atoms together. And when the two atoms are closer than 0.4 nanometers apart, the van der Waals forces repel the two atoms away from each other. If you think of these bonds as being a coil spring with an atom at each end of it, the spring has a state where it is relaxed and is not trying to lengthen or shorten itself. Push the two atoms together a little bit and they try to spring apart. Pull them apart a little bit and they try to pull back together. But since this is a force and there isn't any physical connection between the atoms, I think they are a little weird and sort of magic. In any case, it is these van der Waals forces or bonds that hold the sheets of graphene together to form graphite.

In graphite, for these van der Waals forces to have any effect, the atoms must line up pretty well. They must be within the proper range of distances so the van der Waals forces can act to hold the layers together. There are two forms of this structure in graphite.[1] There is an ABAB structure where every other layer is perfectly aligned. This is what is shown above. The other structure, while much less prevalent, is an ABCABC structure where every third layer is perfectly aligned. The first type—the ABAB type—is called *hexagonal graphite* because it looks like a bunch of hexagonal columns all lined up. The second form is called *rhombohedral graphite* because if you look at how a crystal of this mineral is ordered, you can draw a rhombohedron (mathematical name for a tilted and stretched cube) connecting the three layers. As it turns out, this rhombohedral structure is not thermally stable, and at temperatures above 1200 degrees C it reverts to the hexagonal structure. This is used as a means of purifying naturally occurring graphite[2] for use in all sorts of industrial and medical applications.

In any case, the van der Waals forces holding these layers of graphene together are what make naturally occurring graphite flake easily and are also why pencil lead writes on a piece of paper. Like I noted earlier, the name graphite means, literally, rock that writes. But back to graphite fiber. If you align a series of narrow graphene sheets together in a little round tube you get a graphitic carbon fiber. This is the structure of the short graphite fibers that Roger Bacon of the Union Carbide Parma Technical Center in Cleveland discovered in 1958 (see note 3 for a more information). While Roger Bacon's fibers were not nearly a match for today's high-performance carbon fibers, they marked the beginning of the carbon fiber revolution.

Today carbon fibers that have a graphitic structure are typically made from the thermal decomposition of what is called pitch, which can be from coal tar or from what's left over after crude oil is turned into fuel and lubricants. Pitch is black, sticky, thick stuff with the consistency of wet taffy. The making of carbon fiber begins by pulling a bunch of fibers from this taffy-like pitch using a process much like Singer and Cherry's original taffy-pulling machine (that's why they called it that), and then putting those fibers into a very hot (as high as 2400 degrees F) chamber with an inert atmosphere to drive off everything that isn't carbon. This process is called *pyrolysis* which literally means using fire (heat) to loosen or drive off the volatiles. Since carbon is still solid at temperatures far above where all other molecules and most elements have already boiled off, what remains in the solid fiber is mostly carbon. The process has been improved quite a bit since the original work done by Roger Bacon, Leonard Singer, and Allen Cherry at Union Carbide's Cleveland Parma Technical Center.[3] Where Roger Bacon's fibers were at most 20 percent carbon, and Singer and Cherry's fibers were about a 60–65 percent carbon content, today's ultra-high modulus pitch-based carbon fibers can be as high as 95 percent carbon, which is nearly pure graphene/graphite.

The structure of these fibers is basically a sheet of graphene rolled into a long, very thin solid cylinder to make a continuous fiber with all the benzene rings in the graphene sheet lined up along the length of the fiber. Since graphene is strong and stiff in the plane of the graphene sheet, the resulting carbon fiber is strong and stiff. This type of carbon fiber is called graphitic fiber because it has a graphite/graphene structure with all the 6-carbon rings aligned in the direction that the fiber is pulled. It is also very difficult to break because of the strength of the covalent bonds in these 6-carbon rings. These are the ultra-high modulus, ultra-high strength fibers used in specialized applications. They are far stronger and stiffer than other carbon fibers, but they are also rather brittle. The covalent bonds and the ring structures don't stretch much at all, so they stay in the same form until all the bonds break across the fiber at once. There are other carbon fiber types—one in particular with a turbostratic structure—that are not as strong or as stiff as these graphitic fibers, but they stretch quite a bit more than the graphitic fibers. We will get to these other types of carbon fiber soon.

Graphitic fibers are good conductors of heat and electricity (they have high thermal and electrical conductivity), but only along the fibers and not across the fibers. This is because of the mobility of the electrons in the benzene rings—remember the resonant benzene structure? Since these electrons are free to dance around the rings, they can also dance easily from ring to ring, making graphitic fiber a great conductor of heat and electricity. The electrodes in carbon arc furnaces are made of graphite where the rings are aligned along the length of the electrode. These highly graphitic fibers are used to cool circuit boards, aircraft and truck brakes, missile re-entry nose cones, and satellite structures. They can even be used to house sensitive electronics and act as a lightweight Faraday cage for electromagnetic interference (EMI) and radio frequency (RF) shielding for sensitive electronics.[4]

Pitch itself—the starting material for a lot of the graphitic carbon fibers—is cheap. In the oil industry it used to be thought of as the junk left over after gasoline, diesel, and motor oil were refined out of it. Since the early days of Singer and Cherry's work, quite a bit has changed in the industry in pitch-based carbon fibers. I stated earlier that when Singer and Cherry were doing their original work, two Australian scientists were also working on pitch and discovered this liquid crystal or *mesophase* form of pitch where all the 6-carbon rings were aligned in one direction. These fibers are therefore called mesophase graphitic fibers because of this alignment. What I didn't tell you is how you get to this mesophase. The Australians discovered that if they used a process called melt-spinning, the 6-carbon rings would align themselves and make the pitch fiber non-isotropic. Isotropic means that the properties are the same in all directions, whereas non-isotropic or *anisotropic* means that the properties are different in different directions. This anisotropic nature of all composite fibers makes them somewhat difficult to design with, though it is at the same time extraordinarily useful.

Most mesophase pitch fibers are not completely graphitized or perfectly aligned, because getting to the 95 percent carbon level in a fiber is a lengthy and difficult (expensive) process, so the lower carbon content mesophase pitch fibers—a.k.a. lower quality and lower cost—are the ones used in aircraft and truck brakes and circuit boards where only the electrical and heat conductivity of the fiber are important. The true ultra-high modulus and ultra-high strength carbon fibers that are on the order of 95 percent carbon are used in satellite structures, missile re-entry nose cones, and other extremely high-performance structures where their light weight, high stiffness, high strength for their weight, and high heat and electrical conductivity are critical to the survival of or the operation of whatever they are used in.

The other starting fiber for making carbon fibers is PAN or polyacrylonitrile, which we covered earlier with the history of carbon fiber. First synthesized in Germany before the Second World War and further refined by the Allies after stealing the patents from the Germans, PAN-based carbon fiber became the highest tonnage carbon fiber made today. Toray of Japan makes most of it, either by themselves or licensed to Union Carbide in the United States But I repeat myself.

So what exactly is PAN, how is it made, and how do you make a high strength carbon fiber out of it? To make polyacrylonitrile, first you have to make acrylonitrile, then you have to polymerize the stuff. Acrylonitrile is, at room temperature, a colorless liquid that smells like garlic.[5] It is made up of a vinyl group—an organic group where we have two carbons double bonded together (more about this in the next chapter)—and a nitrile. A nitrile is a group that contains a single carbon atom triple bonded to a nitrogen atom. Acrylonitrile looks like this:

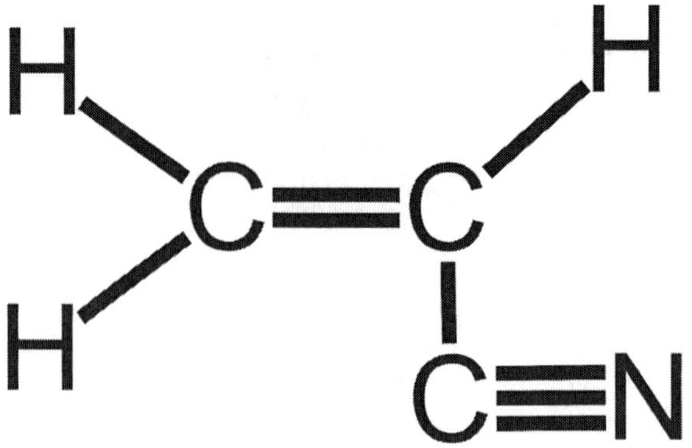

Structure of acrylonitrile (drawing by the author).

The vinyl group is on the left in this picture and the nitrile group is on the right. This is in fact composed of the simplest vinyl group and the simplest nitrile group, so the base molecule of PAN makes a very good fundamental building block for plastics, fibers, etc. When you put a whole bunch of these together, the double bond in the vinyl group goes away and you get a very simple repeating structure whose base looks like the following:

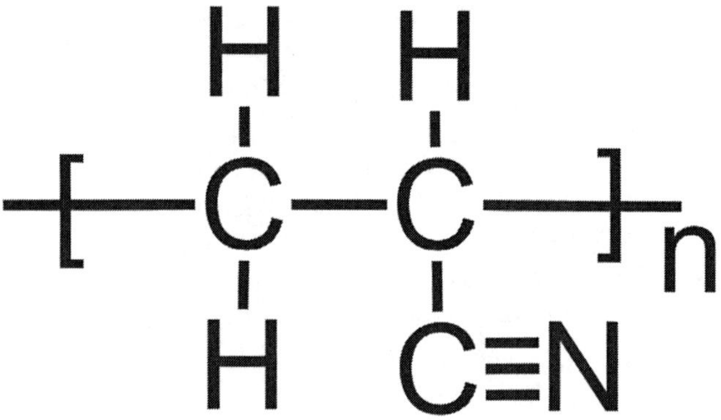

Structure of the repeating acrylonitrile group in polyacrylonitrile (drawing by the author).

Fibers made of PAN have been around a long time, and you may have some of this in your closet. A cloth that is called acrylic is most probably PAN fiber where the PAN has been copolymerized with another organic compound that makes the fiber have the right properties for a sock or sweater. Your "acrylic" socks and sweaters are probably PAN fiber. This was actually the problem with the PAN fiber made in the United States in the early days before the Japanese created Toray carbon fiber. The textile industry was a hot and heavy user of acrylic fiber, but they wanted dyes and other additives put into the fibers so that they could make clothing out of it, which made it completely unsuitable for use as a carbon fiber precursor. In fact, PAN fiber is used today for outdoor items like tents, outdoor awnings, and even sail cloth. Since the researchers at Union Carbide didn't have high quality PAN to work with, they got enamored with rayon fiber instead for their carbon fiber research. The rayon-based carbon fiber was not as strong or as tough as the Toray PAN-based carbon fiber, so eventually they had to give that up. But since the Japanese were still using traditional fabrics (cotton and silk) at the time the rayon work was going on in the United States, their research into PAN was directed toward structural use and creating a high-strength carbon fiber for composites.

PAN is sort of a strange material. It is a thermoplastic, but it doesn't melt. I know this sounds a little weird, and it is, but PAN has some rather different properties than most thermoplastics. It is not soluble in most common organic solvents, but it is soluble in polar solvents, and even in aqueous (water-based) solutions of nitric acid. Strange for an organic material, but that triple-bonded nitrogen makes this a pretty weird thermoplastic indeed.

To make a fiber from this stuff, first you have to start with as pure a form of PAN as you can get. Synthesis of the PAN precursor is done by reacting acrylonitrile with itself in a polar acid bath—sulfuric acid is one of the catalysts that is used to make PAN and there are more depending on who is making it and how they're doing it. All the companies that synthesize PAN have a proprietary formula that is kept as a trade secret and very closely held as intellectual property—this is a very competitive business. In any case, the reaction goes something like the following:

Synthesis of PAN from acrylonitrile (drawing by the author).

The reason strong acids are used in this synthesis is that they tend to produce what are called *free radicals*. Free radicals are unstable molecules that have an unpaired electron. The -OH or hydroxyl group is the best example of a free radical. They are very reactive and very short lived, and they make good catalysts for this type of polymerization reaction. A solution of sulfuric acid in water has lots of these free radicals running around, so it's a pretty good catalyst for this reaction.

What gets produced for use in making PAN-based carbon fiber is a high molecular weight PAN precursor. PAN in its dried and ready-to-polymerize state is a white powder, so what has to happen is that the white powder needs to be dissolved in something that will dissolve it, and then a fiber can be made out of it by a process called *wet spinning*.[6] The PAN is first dissolved in something like dimethylsulfoxone (DMSO) or some other polar solvent, with the resultant being about the consistency of maple syrup. Then this syrupy stuff is extruded through a die with tiny holes in it that are the diameter that the fiber should be, and then it is pulled, washed, and stretched until it solidifies into a fiber. The stretching of the newly made PAN fiber is very important because it aligns the carbon atoms in the polymer so that when cooked at very high temperature to remove everything but the carbon, the carbon-carbon bonds will be aligned along the axis of the fiber. When the fiber is made, thousands of fibers are extruded all at once to make a fiber *tow*.

Now that we have a PAN fiber tow, it needs to be turned in to a carbon fiber tow. The first step in this process is to oxidize the fiber tow to start the cross-linking of the PAN chains. This is a time-consuming process, and the fiber tows are passed through specialized ovens that control the oxidation rate as well as the temperature of the fiber. Since oxidation of the fiber is exothermic (gives off its own heat), there is the real possibility of the fiber catching on fire during this process, so this is done carefully under tightly controlled conditions. But the resulting oxidized PAN fiber has a structure that we are all too familiar with.

See the hexagonal structure here? But one of the atoms in this 6-sided ring is a nitrogen atom because this came from polyacrylonitrile with its triple-bonded nitrogen. By now you can probably see where this is going. We're back to this 6-sided structure that's mostly carbon, so this stuff should make a pretty good carbon fiber. And it does.

These oxidized PAN fibers are actually very useful all by themselves, without being turned into carbon fiber for composites. They are resistant to more oxidation, so they are good as flame resistant fabrics because they don't burn easily—at least not right away. In fact, PAN fiber makes one of the best flame- and heat-resistant fabrics that exists, even more so than Nomex®.

Oxidized and stabilized PAN fiber (from Konstantopoulos et al., "Introduction of a Methodology to Enhance the Stabilization Process of PAN Fibers by Modeling and Advanced Characterization," *Materials*, vol. 13, 2020, Figure 1, doi: 10.3390/ma13122749, Creative Commons Attribution License).

Now that we have this oxidized PAN fiber, the next step is to carbonize or pyrolyze it to make a carbon fiber. This is done at high temperature—up to as high as 3000 degrees F—in an inert atmosphere. In fact, some highly graphitized PAN-based carbon fiber is subjected to as high a temperature as 4500 degrees F because at that temperature all that is left is the carbon. And since this entire process is intended to drive off everything but the carbon, the fibers themselves lose about half their weight and shrink by some 10 percent or so in length and more in diameter. The actual ratio of beginning weight to ending weight is about 2:1 for highly graphitized PAN-based carbon fiber.

Now that we have a tow of carbon fiber, what's next? Well, one of the problems with this form of carbon fiber is that it doesn't stick to much of anything. But because adhesion to things is a surface phenomenon, there are good solutions. If there weren't, carbon fiber wouldn't be worth much to us as a material, because we need it to stick to the epoxy resin we're going to put it in to make something out of it. The next step, then, is to etch the surface a little bit with things like nitric acid or sodium hypochlorite. As this fiber was developed, researchers found that if they also applied an electric field or electric charge to the solution, since the fiber is such a good conductor of electricity, it would etch more evenly, and they got a much better fiber surface preparation. Next comes something called *sizing*, which is a proprietary coating on the fiber that makes it easier to handle and protects the fiber from all the handling processes it has to go through—like weaving into fabrics or bundling into tows with more fibers than the original tow. These sizings are also highly proprietary and vary for different fiber uses. For example, the sizing for single, large-tow fiber is slightly different than the one used if the fiber is going to be woven into fabric. This results in a whole bevy of options for the composites designer to choose from when deciding what fiber to use for a particular application.

But this is both the beauty and power of this material, as well as its Achilles heel. The beauty and opportunity that composites presents is that there is an infinite number of potential combinations and permutations available to the designer to use. The challenge for the designer is that there is an infinite number of potential combinations and permutations available—it can be difficult to make choices. But we will get to why the good composites designer has to be a good systems engineer when we get to the chapter on composites design and analysis.

Glass Fiber

Enough about carbon fiber, what about the most common form of composite fiber used in the industry today—glass fiber? We learned quite a bit about what glass fiber is in previous chapters, so now we have to delve a little deeper into how it's made, how it's processed, how the chemistry and crystal structure of different glass fibers are different, and what that means for the composites designer or engineer.

There are many types of glass fibers, each of which is mostly silica with other elements sprinkled into the mix that change the properties of the silica in useful ways. Pure silica fiber is very difficult to make and work with because it needs to be heated to 1200°C just to soften it enough so that the fiber manufacturers can extrude it. And silica fiber by itself is more brittle than fibers that have other additives that go into the glass melt itself. We learned earlier that the original glass fiber made by Games Slayter was a soda-lime glass—meaning that it was mostly silica with added sodium carbonate (Na_2CO3) a.k.a. soda ash, and calcium oxide (CaO) a.k.a. quicklime. This type of glass fiber is now known as A-glass for alkali glass because of the addition of the quicklime, which is very alkaline when dissolved in water. It is still one of the most heavily used types of glass fiber because it is the most common one used in fiberglass insulation.

There are other types of glass fibers as well, such as C-glass or low-corrosion glass with much less alkali, E-glass, which we have already touched on, and a stronger and stiffer glass called S-glass (now called S-2 glass), or structural glass. Each of these glasses has differing amounts of metal oxides added to them that modify the mechanical, electrical, corrosion, workability, etc., properties of the glass. A table of the compositions of the most common glass fibers is below.[7]

Percentages of Metal Oxides in Various Glass Fiber Types

Glass fiber	A Type	C Type	D Type	E Type	Advantex	ECR Glass	AR Type	R Type	S-2 Type
Oxide	%	%	%	%	%	%	%	%	%
Silicon dioxides (SiO2)	63–72	64–68	72–75	52–56	59–62	54–62	55–75	56–60	64–66
Alumina (Al2O3)	0–6	3–5	0–1	12–16	12–15	9–15	0–5	23–26	24–26
Boron trioxide (B2O3)	0–6	4–6	21–24	5–10	<0.2	-	0–8	0–0.3	<0.05
Calcium oxide (CaO)	6–10	11–15	0–1	16–25	20–24	17–25	1–10	8–15	0–0.2
Magnesium oxide (MgO)	0–4	2–4	-	0–5	1–4	0–4	-	4–7	9.5–10.3
Zinc oxide (ZnO)	-	-	-	-	-	2–5	-	-	-
Barium oxide (BaO)	-	0–1	-	-	-	-	-	0–0.1	-
Lithium oxide (Li2O)	-	-	-	-	-	-	0–1.5	-	-
Sodium oxide + potassium oxide (Na2O+K2O)	14–16	7–10	0–4	0–2	-	0–2	11–21	0–1	<0.3
Titanium dioxides (TiO2)	0–0.6	-	-	0–0.8	-	0–4	0–12	0–0.25	-
Zirconium dioxides (ZrO2)	-	-	-	-	-	-	1–18	-	-
Iron oxide (Fe2O3)	0–0.5	0.8	0–0.3	0–0.4	-	0–0.8	0–5	0–0.5	0–0.1
Fluorine (F2)	0–0.4	-	-	0–1	-	-	-	0–0.1	-

Note here that E-glass fiber has much more alumina, calcium oxide, and boron and much less sodium and potassium oxides (soda ash) than the A-glass or soda-lime glass used for bottles, windows, and the fiberglass insulation in the walls of your house. And, as mentioned, it is the reduction of soda ash and the addition of alumina and boron oxide to E-glass that gives it its strength, toughness, and durability as well as its electrical insulation—hence the *E* in E-glass. The high soda ash content of A-glass causes it to be brittle, as any kid learns when a baseball sails through their parents' front picture window. E-glass has alumina in it, which makes it tougher and less brittle, but E-glass has more lime in the form of calcium oxide than A-glass, which makes it susceptible to corrosion, especially from chloride ion attack—remember the osmotic blisters on boat hulls? The reduction in the amount of CaO (lime) in C-glass makes it more resistant to corrosion at the expense of lowered strength and toughness, so it is typically used as a protective layer in things like the inside of industrial fiberglass process piping where what's going through the pipe is corrosive. There is also ECR-glass, which is a corrosion-resistant, electrically conductive alternative to E-glass, whereas E-glass is a good electrical insulator. The addition of a little zinc oxide and titanium dioxide to E-glass protects ECR-glass from corrosion and makes it conduct electricity. Both additives are expensive, so ECR-glass comes at a cost premium. It is therefore commonly relegated to special applications that can justify the additional cost and need the extra strength and electrical conductivity of ECR-glass over E-glass.

So how is this stuff made? We learned a bit about the early glass fiber manufacture in the history chapter. Fiberglass insulation, or glass wool as it used to be called, is typically made using a melt-spin process. In this process the molten glass is run into a round, spinning crucible with lots of little holes on the outside surface of the part that feeds the molten glass, for the molten glass to escape in long, string-like fibers. Centrifugal force pushes the molten glass out of these little holes and makes a mess of curled up fibers. It's a lot like making cotton candy only at a much higher temperature, and molten glass doesn't taste as good as cotton candy.

But this process won't work for long, continuous fiber, so glass fibers for composites are made using a different process. There are two different processes for making glass fiber and both use what is called a *bushing plate*,[8] which is basically a small, heated extrusion die with lots of little holes for the fiber to be extruded from. These bushing plates can have between 200 and 4000 holes for extruding fibers, in multiples of 200. The two processes are the marble process and the direct melt process, depending on where the fiber manufacturing facility exists. If the fiber manufacturing facility and the glass-making facility are located in the same place, the newly made glass in its original molten form is put directly into the extruder that has the bushing plate(s), and glass fiber is pulled out the other end in a continuous process. In the marble process, the glass-making happens in one place, where the marbles are made. Then the glass marbles are shipped to the fiber manufacturer where they are re-melted and put into the extrusion machine and extruded through the bushing plate(s) into glass fibers.

There is a lot of process engineering that goes into making sure that the glass is at the right temperature, the extrusion pressure is just right, the little holes or nozzles in the bushing plate are the right shape and size, etc. The extrusion of fiber usually happens vertically with the fiber coming out of the bottom of the bushing plate where all the fibers are grabbed by a device that continuously draws out the fiber. If the fiber is extruded too quickly it will break apart into little, short sections. This is because when the little drops of glass leave the bottom of the bushing plate, they leave a thread of molten glass behind them. This thread becomes the fiber, and as long as the fiber is drawn out at the same rate that the glass comes through all the little nozzles, you get continuous fiber. Timing this process perfectly allows fiber manufacturers to make fibers that are continuous until they run out of molten glass in the marbles or the crucible where the glass is being made.

As the fiber is made, very soon after it comes out of the bushing plate, a coating is put on the outside of the glass fiber called sizing in a very similar manner to how carbon fiber gets a sizing. The sizing for glass fiber is different than that for carbon fiber, but the idea is the same. The sizing protects the fiber as it is wound onto bobbins, makes it work more like a textile fiber for handling and weaving, and also helps the fiber stick well to composite resins.

This whole process happens very quickly in the manufacture of glass fiber—especially in the manufacture of E-glass fiber. The fiber manufacturers have figured out how to make it at a rate of nearly a kilometer a minute. That's fiber coming out of the fiber-making machine at nearly 40 miles per hour. That's how they can make so many tons of the stuff in a year. It is made really, really fast. It's also a reason that it is inexpensive and the fiber of choice for the majority of composites applications where the extra strength and stiffness of carbon fiber is not required. The carbon fiber manufacturers can't make fiber that fast because it must cook in very hot ovens long enough to drive off everything that isn't carbon.

What does this mean for the composite designer? It means that you should start your design investigation using the properties that you can get from glass fiber reinforced plastic—or fiberglass—to find out if this material will work for you. And you need to keep in mind the properties of S-glass, ECR-glass, etc., as you begin your investigation so see if maybe one of those might be a better choice. It all depends on the requirements of the structure and of the application, and the choice of both fiber (string) and the resin system (glue) need to be made based on what they are going to be used for.

Other Fiber Types

There is a myriad of other fiber types available to the composite designer, but they really come down to two classes: organic fibers and specialty fibers. In the organic fibers we have things like the aramid fibers Kevlar, Nomex, Twaron®, and Technora®, ultra-high molecular weight polyethylene such as Spectra, and Dyneema®, and liquid crystal fibers such as Vectran and SIVERAS™.

Specialty fibers are those that are not used as much as the typical composite fibers—glass, carbon, and organic. They include boron fiber, silicon carbide fiber, quartz fiber, and some others in very small quantities. Since each of these specialty fibers is heavier than all the other fiber types, they are only used where their unique properties are needed.

Aramid Fibers

Aramid is short for *aromatic polyamide*. I don't expect you to remember this, but I think that things like this are important to introduce so that even the casual reader knows where to look to find the vernacular used in the industry. There are a number of places throughout the book where I will take an aside and try to explain in simpler terms what these terms mean so that when you're done reading, you will have had them introduced to you and will know where to look to learn more.

The most common and well-known aramid fibers are Nomex and Kevlar. We are going to focus on Kevlar as a fiber for composites because Nomex is used primarily in firefighters' clothing and other applications where its flame resistance and resistance to melting and igniting have made it the fabric of choice.

Kevlar was invented in 1965 by DuPont chemist Stephanie Kwolek,[9] and it has been one of the most successful fibers in DuPont's history. Kevlar-29, the original Kevlar, is poly-paraphenylene terephthalamide, a mouthful at best (again—you don't need to remember this). DuPont had been working with nylon, which is a polyamide plastic material, and making fibers out of nylon for things like rope and fishing line. Stephanie Kwolek was in the group working on polymer chemistry at the time that DuPont was searching for a higher strength fiber than nylon. Her boss at the time, DuPont research fellow Paul Morgan, suggested that putting an aromatic (benzene ring) backbone on the polyamide polymer would make a stronger, stiffer fiber. And since the team knew that the resulting polymer would have a very high melting point, Kwolek worked to find the right solvents to dissolve the reactants in to make a polyaramid. The result was Nomex, chemical name poly(m-phenylene isophthalamide). The *m-* stands for "meta", which denotes the positions of the amine groups on the benzene ring.

The starting amine used to make Nomex is meta-phenylenediamine, which looks much like you would expect (see figure to the right). It is a benzene ring with two amine groups—one at the *R* position in the figure above and one at the meta position on the ring above. What Kwolek discovered when she put together meta-phenyldiamine and isophthaloyl chloride (isophthalic acid with chlorines on each of the acid groups) was a liquid crystal polymer, which is a cloudy white solution of liquid and polymer particulate. When dried, the resulting white powder can be spun into fiber to make Nomex.

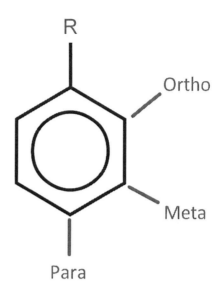

Ortho, meta, and para positions on the benzene ring (drawing by the author).

Nomex is an enormously successful fiber for DuPont because of its flame resistance. When you make a fiber out of it the polymer chains don't completely line up, so the resulting fiber is only about as strong as nylon, but it is very flame and heat resistant. This is where Nomex gets its fame and also its usefulness.

In 1964 in anticipation of the gasoline shortage that spiked fuel prices all over the world, Kwolek and her team started looking for a stronger,

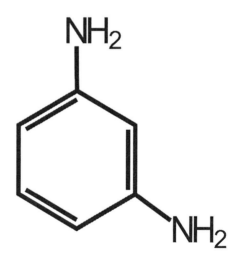

Meta-phenylenediamine—the backbone of Nomex® (drawing by the author).

lighter fiber than steel for use in making car tires. And, since she had been working with the chemistry of Nylon and had figured out what solvents to use to make Nomex, Kwolek thought—rightly so—that if she used the

-para version of phenyldiamine to make a fiber, the liquid crystals might be able to line up correctly and make a strong and stiff fiber. The compound that Stephanie Kwolek used to create Kevlar is shown below:

She was right because what she eventually came up with was Kevlar-29, which started the revolution that aramid fibers have become. A schematic of the reaction that Kwolek discovered is shown below where para-phenylenediamine (on the left in figure below) is reacted with terephthaloyl chloride—another mouthful that I do not expect you to remember. Note the -*para* positions on the benzene ring for both of these compounds. All you have to do is remember where the attachments to the benzene rings are and you will understand why Kevlar is as strong as it is and has such a high stiffness or elastic modulus—all the benzene rings are lined up perfectly.

When this polymer—which is another liquid crystal

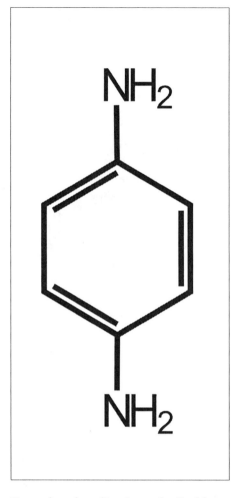

Para-phenylenediamine—the Backbone of Kevlar® (drawing by the author).

Chemistry of the reaction that creates Kevlar® 29 (10) (produced by Roland. chem, Creative Commons CC0 1.0 Universal Public Domain Dedication).

polymer—is mechanically pulled to make a fiber, all the chains orient in the direction of the fiber, making the fiber stronger than steel at a small fraction of the weight of steel.

So here we are back at the benzene ring and fibers again. This is again the reason that understanding, at least at a basic level, the chemistry and physics of carbon and its aromatic or benzene 6-carbon ring structures is so important to understanding composites.

Kevlar fiber maintains its strength and toughness down to very low, *cryogenic* temperatures (like minus 250 to 460 degrees F—"absolute zero" in kelvin) quite unlike most fibers that become brittle when very cold. It also is a very poor conductor of heat—a good insulator. For this reason, it is used extensively in the field of particle physics as a thermal standoff between superconducting magnets that have to be kept at temperatures just above absolute zero. This fiber is also used for structural support where very low heat leakage is required. It also is used for ballistic armor protection—as in Kevlar vests that most police force SWAT teams use. Kevlar is also used extensively in sporting goods, climbing rope, paraglider suspensions, lightweight bicycle tires, and even in Nike basketball shoes.

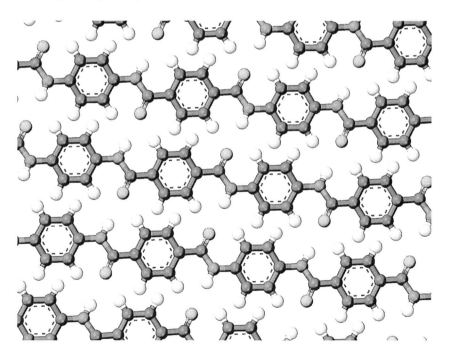

Ball-and-stick model of a single layer of the Kevlar® crystal structure (produced by Ben Mills and Jynto based on designs by Benjah-bmm27, Wikipedia, public domain).

And, of course, Kevlar is used in composites, commonly in combination with glass or carbon fibers as an abrasion- and corrosion-resistant outer layer over a glass or carbon epoxy structure. The Ferrari F40 made extensive use of Kevlar and carbon fiber composites married with aluminum for structural body panels. And Kevlar is used as a primary structural fiber in several high-performance aerospace structures—even in space. The Mars Pathfinder used Kevlar fiber-reinforced air cushions to land on Mars.

Shortly after DuPont brought Stephanie Kwolek's invention of Kevlar to the market in the United States in 1971, a Dutch company by the name of AzkoNobel developed a para-aramid fiber nearly identical in both chemistry and properties to Kevlar. They had originally called this fiber Arenka. In the late 1960s and early 1970s when this fiber was first developed, AzkoNobel was having financial difficulties, so it took them until 1977 to put it into production at a small scale. Then in 1984, 13 years after the introduction of Kevlar to the market, the fiber was renamed Twaron, and in 1987, 16 years after the introduction of Kevlar, Twaron was introduced as a commercial product.[10] Finally, in 1989 the aramid fiber business unit of AzkoNobel was sold to the Japanese industrial conglomerate Teijin Group which is now called Teijin Twaron BV headquartered in Amsterdam. In 2007 Teijin dropped the name Twaron and now the fiber is called Teijin Aramid.

Twaron fiber is made using the same starting materials as Kevlar and results in the same chemistry. However, the process for spinning Twaron fiber is less complex than the process for making Kevlar, so this fiber is competitive with DuPont's Kevlar. Just like Kevlar, Twaron is the same polymer as Kevlar and is made by the same reaction as Kevlar. What is different is that instead of using DuPont's difficult and expensive process for spinning the liquid crystal polymer into a fiber, they used an initially solid solution of the powdered polymer and frozen anhydrous (without water) sulfuric acid in powder form to dissolve the polymer. Actually, the two powders are mixed together cold and then very gently heated until the mixture becomes very viscous—sort of like a cross between maple syrup and wet taffy. We know what happens next. This syrupy, taffy-like liquid is forced through the melt-spin machine in the same manner as nylon fiber is made. This process for making Twaron fiber was patented by AzkoNobel so that the commercial introduction of Twaron could occur, and AzkoNobel wouldn't violate any of DuPont's patents.

As is common in the history of composites, there was a reason that Teijin Group bought the Twaron business unit from AzkoNobel. In the mid–1970s Teijin was experimenting with aramid fibers to make a more heat-resistant and abrasion-resistant fiber than Kevlar. They used a combination of two different amines—para-phenyldiamine and 3,4'-diaminodiphenyl ether. This second amine is made up of two benzene rings connected by an oxygen atom with amine groups at each end. But one of the amine groups is clocked off the center of its benzene ring.

When Teijin made fiber from this mix of precursors, they found that the fiber had excellent mechanical and thermal properties, similar to Kevlar and Twaron, but the process for making it is much simpler. Rather than having to use a solid solution of the polymer and anhydrous sulfuric acid, fiber can be spun directly from the initial polymerization reaction. Teijin called this fiber Technora, and it is used extensively to reinforce rubber hoses and belts, for armor cladding around optical cables, for windsurfer and hang glider sails, and was even used as one of the materials for the parachute that allowed *Perseverance* to land on Mars. It is not, however, used much in composites.

Structure of 3,4'-diaminodiphenyl ether (drawing by the author).

Other Organic Fibers

Most of us have heard of a fiber called Vectran. This is another liquid crystal polymer, but it is not an amide-based polymer. Instead, it is an aromatic polyester.[11] Since it is in this same class of fibers made from liquid crystal polymers, the reaction that makes the Vectran polymer is a condensation reaction between two aromatic compounds: 4-hydroxybenzoic acid and 6-hydroxynaphthalene-2-carboxylic acid. Again, here we go with the mouthful that you get with this organic witch's brew style of chemistry, but by now you should be familiar with all the pieces of the names of these two chemicals to be able to figure out what their structure looks like. The exception would be the naphthalene that forms the backbone of Vectran. Naphthalene is the simplest of the poly-aromatic compounds and is basically two benzene rings stuck together. This stuff is also called white tar, camphor tar, etc., and is the primary component of moth balls.

When you make an acid out of this stuff by adding carboxylic acid

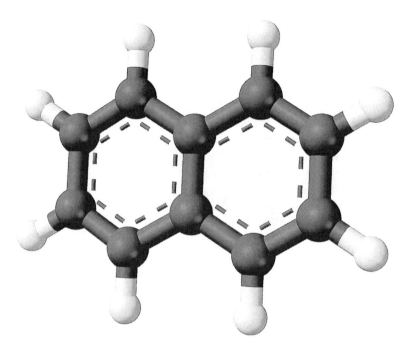

Structure of naphthalene (produced by Benjah-bmm27, Wikipedia, public domain).

groups to it, marry it up with hydroxybenzoic acid in another condensation reaction, you get the Vectran polymer.

One of the major differences between Vectran and the original aramid fibers is that this liquid crystal polymer will melt, so the fiber can be melt-spun like nylon. In the melt-spinning process, the molecules will line up, and the resulting fiber has well-aligned polymer chains, so it is less expensive to make than Kevlar or Twaron while maintaining similar strength and stiffness to Kevlar. It can be used for nearly the same

Vectran® structure (drawing by the author).

products that are being made from Kevlar today. It does fray more than Kevlar, so it is typically coated with a polyurethane coating before use, which makes it somewhat heavier than Kevlar. Vectran also has low moisture absorption, high heat stability, good dimensional stability, and good resistance to most chemicals.

Vectran was originally developed by Celanese Corporation's advanced materials business unit in South Carolina. Since then, Celanese sold its Vectran business to Kuraray Co. Ltd, a large Japanese chemical company with their main manufacturing facilities in Kurashiki, Okayama, which started production of Vectran in 1990. By 2007 Kuraray became the sole producer of Vectran worldwide.

The last group of fibers in this organic fiber mix are the ultra-high molecular weight polyethylene fibers—Spectra and Dyneema are the best examples. These fibers are classed as polyolefin organic compounds which are polymers made from simple alkenes. In our case, the original organic compound that makes polyethylene is ethylene which is two carbon atoms double bonded together with two hydrogens attached to each carbon. We touched on ethylene a while back, so this should be familiar to you. And it will come as no surprise to you what the structure of polyethylene looks like.

Polyethylene is one of the most common plastics used today. Your grocery bags are polyethylene, as is the plastic sheeting that you put down on your floor when you paint the walls of your house. It is simple to make and very inexpensive. But if you want to make a structural fiber out of this stuff, you have to have very long chains of ethylene, so for the ultra-high

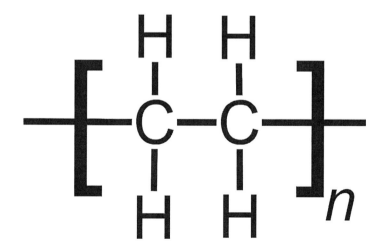

Structure of polyethylene (drawing by the author).

molecular weight polyethylene fibers, the n in the figure above has to be higher than 100,000. And if you can align these monstrous-length molecules into a fiber, the fiber is quite strong. These fibers do, however, have rather low heat resistance, and tend to melt, whereas the aramid fibers don't.

Spectra and Dyneema, the two best known of these fibers, have tensile strengths comparable to high-strength steel. But since the density of polyethylene is so low, they weigh about an eighth of a comparable high-strength steel. They are used where high strength and low density are needed and where they will not be subjected to heat. The fibers are nearly impervious to moisture, have high chemical resistance, and especially high abrasion resistance. This makes them useful as an outer layer over a fiberglass or carbon fiber composite structure where resistance to abrasion and low water absorption is needed. And since they are very low weight for their strength, they have been used in skis, snowboards, and several other sporting goods and other applications, mostly in combination with carbon and glass fiber. And, since polyethylene is so slippery, making a composite material with it requires plasma treatment or plasma etching of the surface of the fiber to make the epoxy stick.

One more organic fiber to learn about before we move on. There is a polypropylene (three carbon groups rather than two) fiber that we need to mention. Again, this is a high molecular weight, highly oriented polyolefin fiber like ultra-high molecular weight polyethylene fiber. The fiber was developed by a small business in South Carolina called Innegra, and they have branded their fiber Innegra™ S. The fiber was commercialized in 2012 after being developed using support from the U.S. Government's Small Business Innovation Research (SBIR) program. This is one of the success stories that has come from the SBIR program, and Innegra™ S has broadened its scope and manufacturing capabilities since, and even formed a partnership with BGF Industries, a composite fiber fabric manufacturer in Virginia. Innegra™ S is the lightest weight high-strength fiber available today.

Boron Fiber

Now that we have covered most of the organic fibers, there is one more class of fibers that I need to introduce—ceramic and other hard material fibers. The most common of these types of higher density but higher strength and incredibly heat resistant fibers are boron fiber and silicon carbide fiber. The first fiber material produced for composites for aerospace applications was boron fiber. To make a fiber out of Boron you need a backbone of some other high-strength material. Boron fiber is basically elemental boron deposited on a thin tungsten wire using a chemical vapor deposition (CVD) process. This process must be done at very high temperatures, which is why the inventors of this fiber used tungsten cores. It was Texaco in 1959 that demonstrated the ability to make a continuous, high-strength, high-modulus fiber using their CVD process. This got the attention of the U.S. Air Force Materials Lab, which provided funding to scale up Texaco's process, and boron fiber was commercially available by 1964.[12]

Boron fiber is not as light as carbon fiber because tungsten itself is heavier than lead, but the fiber has a better strength-to-weight and stiffness-to-weight ratio than almost all metals. And, unlike most composite fibers, boron fiber has very high compressive strength, primarily because the tungsten core makes for a much larger diameter fiber than carbon or other high-performance fibers. A thin boron fiber composite is much less susceptible to buckling under compression, which makes it a good choice for wing and tail skins of lightweight, high-performance aircraft. Boron fiber was used in both the F-14 Tomcat and the F-15 Eagle fighter aircraft for horizontal and vertical tail skins to reduce flutter without increasing the weight of the aircraft, especially since the tail skins are in the rear of the plane. This allowed for a larger control surface that would not flutter. That is one reason that both the F-14 and F-15 have proven to be such amazing aircraft. When the F-14 and F-15 were in development, continuous filament carbon fiber was either unavailable or prohibitively expensive, so boron fiber was used instead. Boron fiber was also used in the original space shuttles primarily for its compressive strength. The first use of boron fiber in fighter aircraft happened in 1969, before the carbon fiber revolution. More recent fighter aircraft like the F-16 and F-18 used carbon fiber epoxy for their tail skins.

Today boron fiber is used mostly in high-performance prepreg tapes where it can be used to reinforce an otherwise mostly glass composite structure. It has also been used to stiffen golf club shafts, tennis rackets, hockey sticks, etc., where its high tensile and compressive strength and high stiffness are used in specific areas to enhance performance at a relatively low cost penalty. And it has been used in hybrid applications with a carbon fiber composite for those areas of a structure that come under high compressive load.

Silicon Carbide Fiber

Finally, the last fiber I need to mention here is silicon carbide fiber—yes, they make fibers from the stuff embedded in the teeth of your circular saw blade. Silicon carbide (SiC) is the second hardest substance known to occur naturally, although the mineral form of SiC, Moissanite—again, a mineral named after its discoverer—is pretty rare. Moissanite was discovered by the French chemist Henri Moissan in 1893.[13] It is mostly made industrially and has been in production under the name Carborundum since 1890 when it was put into production by Edward Goodrich Acheson.[14] Acheson patented the method for making SiC in 1893 and formed the Carborundum Company to make abrasives out of it.

Making a fiber out of SiC is somewhat difficult: there are three methods for making it that are used today.[15] The one that has been in use the longest was invented in 1975 and is called the Yajima process.[16] A polymer that is not yet a ceramic but has the right mix of silicon and carbon is spun into fibers using the typical melt-spin process. The resulting fiber, since it is going to become a ceramic, is called a *green* fiber. This is common nomenclature for the ceramics industry, and it means that the fiber as it comes out of the spinneret still needs to be fired at a high temperature to become a ceramic. The green fiber goes through several steps in the firing process to remove almost everything that isn't SiC to make a SiC fiber.

The second method for making this fiber is very similar to the way that boron fiber is made, only the thin fiber substrate that the SiC is deposited on using CVD is usually a carbon fiber. This makes a composite itself where the two materials—carbon and SiC—are combined to make something with different properties than either of the respective starting materials. This fiber is marketed as the SCS fiber and has been used to reinforce ceramics and metals to make ceramic fiber–reinforced composites and metal matrix composite materials. It is the material of choice for the newer GE turbofan jet engine turbine blades because of its very high strength and stiffness and extraordinarily high heat resistance.

A third method of making SiC fiber, which is new, is similar to the way SCS fibers are made, but there is no carbon core. Instead, the SiC is deposited using a laser-assisted CVD process (LCVD) directly into a fiber with no core. A small company called Free Form Fibers in Saratoga Springs, NY, developed the LCVD process and makes a number of LCVD-based ceramic fibers. They began developing the LCVD process in 2009 and, with significant SBIR funding, formed a company to develop these fibers. Recently they raised $2.5M in venture funding to complete the commercialization of their process. This is another SBIR success story where their funding, along with research grants from NASA, the U.S. Army, and NSF enabled the company to develop a lot of the particulars of the LCVD process and to commercialize and produce these high-performance fibers for the aerospace industry.

That's about it for the string chapter of this book. Now it's time to learn about the glue that binds them together, so we need to move on.

The Glue—
a.k.a. Composite Resins

This is the next major branch of our semantic tree of composites: the glue. Remember Bakelite? And remember the structure of phenol, or benzene alcohol? This invention of Leo Baekeland's started the plastics industry and is the foundation for nearly all composite resins that are produced today. Phenolic resins themselves are relatively simple polymers, so let's start there and work our way up through polyesters, vinyl esters, and finally, epoxies.

Phenolic Resins

First, we have to remember what phenol looks like and the importance of its basic ring structure. The figure below is a reminder of this very simple compound that has so many wonderful properties and applications.

It is this ring structure—the benzene ring—that really creates all the magic in phenolic resins, epoxy resins, polyester resins, vinyl ester resins, you get the drift. That's why we spent so much time covering benzene and its ring structure.

When you react this alcohol with formaldehyde (yes, the toxic smelly stuff we all know and hate) using the right conditions, it forms a hard and tough polymer called a phenolic resin.

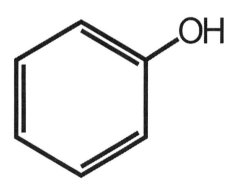

Chemical structure of phenol or benzene alcohol (drawing by the author).

So just to get this party started, here's the chemical structure of formaldehyde.

Formaldehyde by itself is a gas that is very reactive and does not last long when in the open air because it will react with almost anything. But, when formaldehyde is mixed with water, it stays in solution and ends up in equilibrium with a very similar molecule called methylene glycol. But we will get to this reaction in a minute.

First, when you mix these two compounds (formaldehyde and phenol) together with either a water-based acidic or basic

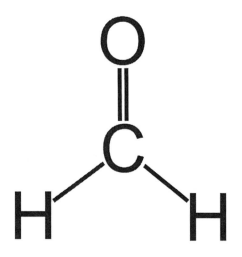

Chemical structure of formaldehyde (drawing by the author).

solution to catalyze the reaction, you get the backbone of phenolic resins. If you remember what I wrote about hydrogen and acid-base reactions, you know that what this means is that in acid catalyzation the acid is a proton donor, and in a base-catalyzed reaction the base is a proton acceptor. And what is a proton but a hydrogen nucleus? Again, it's hydrogen that is the actor here in the chemistry of these reactions, and hydrogen that rules the day in the chemistry of composite resins. Our good friend water is the vehicle that allows for free protons to make acids and bases what they are.

What does this reaction look like? As it turns out you get very different resins if you use a base catalyst versus an acid catalyst. And this is because the acid catalyst causes a much different chain of reactions than does the base catalyst. An acid catalyst is used to make what are called Novolac resins (Leo Baekeland's first resin), whereas the base catalyzed reactions make what are called *resole* resins. The synthesis of these resins is a multi-stage process where phenol and formaldehyde are mixed in water and a catalyst is added. This is because when formaldehyde is in a water solution—which is how it's made and sold—it exists in equilibrium with methylene glycol, and it is the methylene glycol that reacts with phenol to form the backbone of the phenolic resin. See, we got back here and now we're going to learn about formaldehyde-methylene glycol. But I had to give you some background about the acid- and base-catalyzed reactions so you would understand why this is important.

When you react methylene glycol with phenol in an environment where there are plenty of hydrogen nuclei around, like in an acidic water

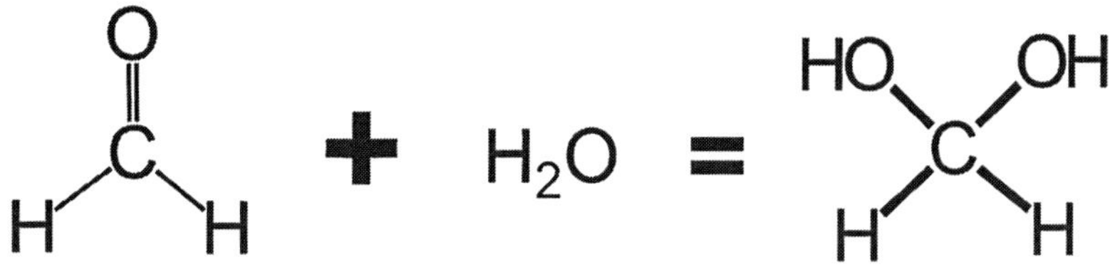

Formaldehyde + water in equilibrium with methylene glycol (drawing by the author).

solution (protons—remember acids are proton donors), you get what is called a Novolac resin plus a water molecule. This is a mixture of different forms of the same compound, but the new addition to the phenol molecule attaches in two different places on the benzene ring. An ortho attachment is where the new addition is right next to phenol's OH group, and a para attachment occurs where the new addition is on the opposite side of the ring from the OH group.

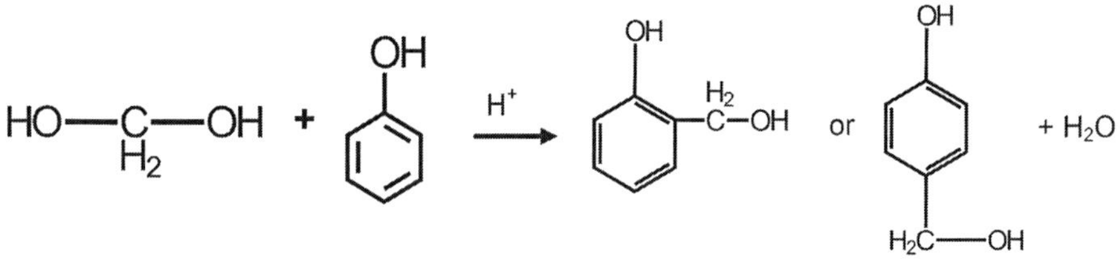

Ortho (right) and para (left) Novolac precursor synthesis (drawing by the author).

Note that the new additions to the phenol molecule retain that proton and make these resins very reactive and want to bond together. And when they do bond, they make a complex mixture of ortho-ortho, para-para, and ortho-para complexes.

Then, when you react these together you get a very complex and twisted polymer that becomes a three-dimensional mess of benzene rings all stuck together.

Note here where the OH groups are as well as where the little asterisks are—these are the reaction sites or the places where more of the ortho-ortho, para-para, and ortho-para Novolac precursors can attach to make what is called a cross-linked polymer. The form shown above is how the resin is provided to the user, a form that has all these little reaction sites on it that are ready to be reacted into a solid.

Three different configurations of base Novolac resin molecule (drawing by the author).

Novolac resin structure (drawing by the author).

This form of the resin—the one with little reaction sites ready to be bonded to still more of the resin—is delivered to the user as a liquid ready to be made into a solid. This is what you would buy off the shelf at your local phenolic resin store (sorry, had to do that—these resins are readily available online). This form still needs to be cross-linked to make a solid resin product. One of the common cross-linking agents is called hexamethylenetetramine,[1] a.k.a. hexa, or hexamine, or just simply HMTA. I know this is a mouthful, but you don't need to remember what this is, you only need to remember what it does. Hexa provides free formaldehyde to the reaction sites (the little asterisks in the figure above) without liberating much of the formaldehyde during the high temperature cure that the resin must go through. And because the Novolac is as complex a structure as you see above, and since these molecules are not flat, but each bond is at a three-dimensional angle compared to the adjacent bonds, this resin forms a 3D network of benzene rings and therefore a 3D network of tightly bonded, tough, heat-resistant, solvent-resistant, and basically permanent material. When Leo Baekeland first started experimenting with this stuff (he was really looking for a replacement for shellac), he ended up using it as a binding agent to bind wood fibers together, thus making the very first organic composite, which he called Bakelite (see, I got you back here didn't I?). As it turns out Leo's shellac was a failure, but his phenolic resin and Bakelite were an enormous success. That's the sign of a real and brilliant inventor—to create something new, important, and commercially viable out of the ashes of failure.

Novolac resins, since they require cross-linking, typically require temperatures higher than room temperature to cure into a solid. The pure resin is a solid and is provided as a powder. To be used as a laminating resin for composites, the solid powder Novolac is dissolved in enough organic solvent to create a liquid with the viscosity needed for how the resin is being used. Pure resin powders are typically used as casting resins, but when dissolved in organic solvents with controlled viscosity, they can be used as laminating resins for composites. When the prepared laminate is to be cured, the entire structure, with all the tooling and whatever mold the laminate is built on to make the desired shape is placed in an oven and cured. This process drives off the organic solvent, and it also drives off the water and free formaldehyde that is produced when the Novolac is polymerized into a solid.

But what happens when you react phenol with formaldehyde in the presence of a proton acceptor—a base? Well, you get what is called a resole resin. These resins are created through a slightly different pathway but end up with a similar result. The reaction is still phenol with methylene glycol, but in the presence of a proton acceptor, the result of reacting these two together is called methylol phenol. Note all the *ols* in this name—that must mean that this is an organic alcohol, which it is. Here's a pic of the reaction that makes this alcohol:

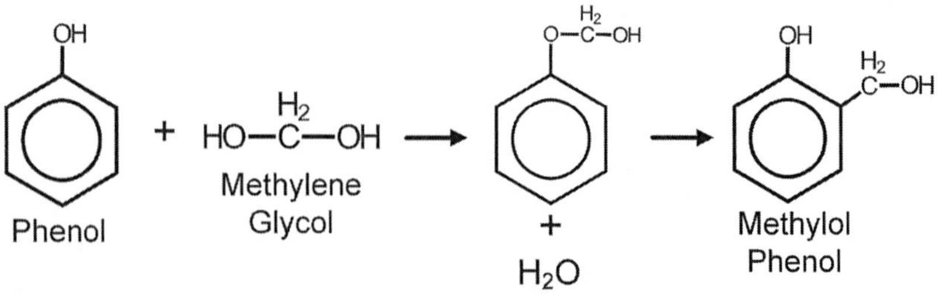

Synthesis of resole resin—phenol + methylene glycol = methylol phenol (drawing by the author).

Since this is an organic alcohol, it can react in more ways with both itself and with phenol to make a variety of different resin backbones. If you react methylol phenol with itself, it can form both a longer chain methylol phenolic, or if there is an oxygen that remains between the two methylol glycol molecules when you stick them together you get dibenzyl ether. Or if you react methylol phenol with phenol—as long as there is formaldehyde present—you get a methylene bridge between two phenols. These three reactions are shown in the figure below:

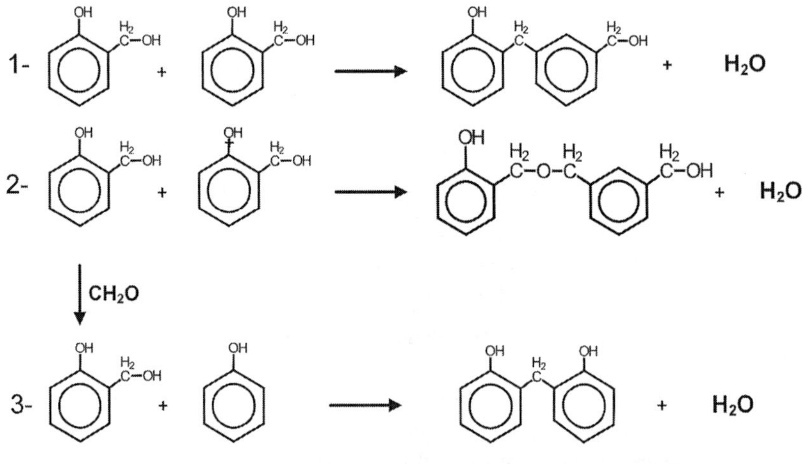

1: Longer chain methylol phenolic; 2: dibenzyl ether; and 3: methylene bridge between two phenols (drawing by the author).

Since this basic reaction can create these three very different compounds, all of which have two phenols attached together, resole resins can have quite a bit if variety. And since the first two compounds, the longer chain methylol phenolic and the dibenzyl ether, can react easily with themselves or with one another without having a cross-linking agent, they will continue the polymerization process as long as there is formaldehyde present. This makes them what are commonly known as *single-stage* or *one-step* phenolics versus the acid solution-formed Novolac resins that require an additional cross-linking agent (curing agent) to turn them into solid. Since these resole resins are alcohols (with the -OH groups), they are slightly polar and almost all of them will dissolve in water at least a little bit.

Because these resole phenolic resins react with one another continuously, when they are supplied to the user they come with a shelf life. Quite a few of the early resole resins had to be refrigerated before use because at room temperature they would continue to react and try to form a solid. Because of this, resole resins were notoriously difficult to use in a continuous manufacturing or production line. If they are still cold, they are very viscous and difficult to work through nozzles and other automated resin supply equipment. And if you heat them up, the reaction goes much faster, so they have limited *pot life*, meaning that once they come out of the refrigerator and you warm them up, you better use them for whatever you're doing right away before they cure and become hard.

One issue that has caused concern is that both Novolac resins and resole resins give off formaldehyde when they cure. That formaldehyde is toxic to the workers in the factory and corrosive to most manufacturing equipment—this is especially true of Novolac resins because of the acidic nature of the resin itself. Plywood and particle board for home construction is typically manufactured using phenolic resins to bind the layers of wood together, or in the case of particle board or oriented strand board (OSB), used to bind the wood chips together to make a solid board. This is because phenolics are not only well suited to making plywood, but they are also the least expensive of all the resins and adhesives available to industry. This is why plywood, particle board, and OSB have the odor they do when new—the phenolic resins are still giving off formaldehyde as they continue curing. This has caused concern with the home construction industry and is why new homes need to have a ventilation system that permits many air changes per day. If the air in your new two story house doesn't get refreshed regularly, the formaldehyde given off by the building materials will expose you to formaldehyde. Not a good thing.

Quite a bit of research has gone into how to either remove or at least reduce the formaldehyde used and produced in phenolic resin synthesis. Some newer resins that have emerged on the market today, primarily to be used in the construction industry, use bio-based polymerizing agents that eliminate the formaldehyde. One bio-based resin is phenol-hydroxymethylfurfural (PHMF).[2] This resin is produced by reacting phenol with glucose, and when both the glucose and the phenol are bio-based, it is a completely bio-based phenolic. In a review by Catherine Thoma et. al.,[3] they cite work going back to 1926 in hydroxymethylfurfural (HMF) to be used as a binder in phenol-based adhesives. Not much progress occurred until the 1970s and 1980s when researchers started investigating using acid dehydration of glucose—basically removal of water from this simple sugar—to form HMF. HMF is seen as a potential inexpensive, bio-based alternative to petroleum-based precursors used in adhesives. This is especially true for the wood industry, which is the largest user of adhesives, accounting for nearly 65 percent of all adhesives used in industrial production today. This is primarily for the construction industry, which uses more wood than any other industry in the world. Coming up with an inexpensive, bio-based adhesive without formaldehyde that could be created using the byproducts of wood processing is an industry game changer and drove quite a bit of work toward making these adhesives and resins at an industrial scale.

For composites, Yuan, Zhang, and Xu[4] in 2014 wrote up the results of a study where they reacted bio-based phenol with HMF to make PHMF as a resin to use in glass fiber composites. Then the put their new "fiberglass" composite through a series of mechanical, thermal, chemical, etc., tests to find out if PHMF performed as well as other phenolic resins to make glass fiber-reinforced composites. What they reported was that this resin performs as well as, if not better than, traditional Novolac or resole resins in this composite material. It is only a matter of time, then, before the remainder of the wood products and composites industries replace the traditional use of phenol-formaldehyde resins with entirely bio-based PHMF phenolic resins.

Polyester Resins

First, what's an *ester*? And why does this ester make polyester resin? According to *Wikipedia*, an organic ester is a compound formed from an organic acid where the hydrogen on at least one -OH group is replaced with an alkyl (hydrocarbon missing a hydrogen) group. The basic structure looks like the following:

The R's in this figure are any hydrocarbon group—typically denoted by R because they can represent typical hydrocarbons like methane,

ethane, propane, etc., or aromatic groups—the benzene ring. Note also that the *R'* attached to the oxygen atom is where the original hydrogen was missing in the organic acid. This group is called an ester group when it is attached to a hydrocarbon.

As it turns out, the most common ester used in polyester resins is made by the esterification of phthalic acid. And it's the phthalic resin that holds the benzene ring that gives this

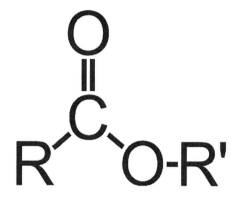

Basic structure of an organic ester (drawing by the author),

resin its strength and stiffness when it is cured into a solid plastic. There are two different phthalic acid *isomers* that are used to make these resins: orthophthalic acid (or just phthalic acid) and isophthalic acid. Note the two groups on each isomer that are ready to give up their hydrogen atoms to form the esters. These groups—called carboxyl groups—are what make organic acids proton donors. You will need to remember the carboxyl group when we get to vinyl esters and epoxies because that's where all the chemistry magic happens.

If you remember the ortho and para positioning of the connections of the benzene rings in phenolic resins, now we introduce the iso positioning. And this makes up all the possible positions that a pair of hydrocarbon groups can take on the benzene ring. Ortho positioning occurs where the two groups are right next to each other, iso where they are separated by one carbon atom on the ring, and para where they are on opposite

Isophthalic (left) and orthophthalic or phthalic (right) acids (drawing by the author).

sides of the benzene ring. This is, in fact, the definition of isomer, which is where a compound has the same chemical formula but takes on a different shape. In most of chemistry this isn't that important, but when it comes to the benzene ring and aromatic hydrocarbons, this positioning of the groups on the benzene ring is critical because when you make something out of one particular isomer, it ends up having properties distinct from that made of a different isomer. This difference is very important in composite resins, and it shows up first in these two isomers of phthalic acid and the different types of polyester resins they make.

In any case, this phthalic acid—whether in its ortho or iso configuration—is the backbone of all polyester resin systems. Esters of these two isomers are easily made by replacement of the hydrogens on the -OH groups of these acids with either more of these acids or with some other hydrocarbon groups. In polyester resins, these are what are called the reaction sites where the resins can be cross-linked (remember phenolic cross-linking?) to form a solid.

Orthophthalic polyester resin is the most common form of polyester resin used in nearly all fiberglass commercial products: boat hulls, bathtubs, tennis rackets, process piping in the chemical industry, non-slip floor grating in factories and meat packing plants, and wherever we need a lightweight, inexpensive material that cleans easily, lasts a long time, and is impervious to most chemicals. It is the material of choice for housings on most common kitchen and bath appliances because it is nearly impervious to water and detergents and can be cleaned and sanitized very easily.

The most common method for making orthophthalic polyester resin is to first create phthalic anhydride from orthophthalic acid by removing a water molecule from it. It is then reacted with maleic anhydride, which is another organic acid with two ester-making carboxyl groups, and propylene glycol—an organic alcohol with two OH groups.[5] These three are reacted together in a condensation reaction (removal of a water molecule) to make what is called an *unsaturated polyester resin*. It is unsaturated because it still has plenty of these reaction sites from all the ester-making groups that still have their double bonded oxygen atom—ready to react and break that double bond.

I hope you're still with me because here's where it gets a bit more interesting. To make a hard plastic from these liquid resins, another source of a benzene ring needs to be added to the mix to cross-link these unsaturated resins together. This is where something that we have all heard of comes into the picture—styrene. We all know about styrene, right? It's the stuff that makes the rigid foam packaging you get when you buy a new toaster. It's cheap, lightweight, can be easily made into a foam product that is mostly small air bubbles and little else, so it makes great packaging. It is also what makes up your yogurt cups, cottage cheese containers, etc. It is used very heavily in the food packaging industry because it is chemically inert and biologically completely inactive. In other words, whatever is put into the styrene container will have no effect on the container and the container will have no effect on what's put in it. It is unfortunately, not bio-degradable, and needs to go in the recycling container. In any case, the chemical structure of styrene is shown below.

By now this sort of thing should look very familiar to you because this aromatic benzene ring-based set of organic compounds basically forms all the glues used in composite materials. In any case, the styrene is used to cross-link between the unsaturated orthophthalic polyester resin and more of the same unsaturated orthophthalic polyester resin—and on and on until there is a hard plastic. When this resin is combined with glass fibers in whatever form the fibers need to be

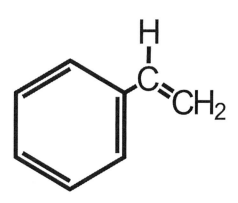

Chemical structure of styrene (drawing by the author).

to make the intended product, and the styrene cross-linking is made to take place by curing the resin, you get a fiberglass part like a boat hull or a bathtub. A run-through of this set of reactions is shown below.

Take particular note of the two styrene molecules situated between these two orthophthalic polyester chains. This is what allows the unsaturated polyester resin to form a three-dimensional structure. Remember that these carbon-carbon bonds can have any angle between them, and when flat benzene rings are situated between two chains, the chains can be at any orientation to each other or to the benzene rings. In reality, this chain is never flat, but is entirely three-dimensional. It's tough to illustrate that on a flat sheet of paper, so you have to use your imagination to imagine what it would look like when the chains are twisted not only within the chain, but also with relation to each other and with relation to the two benzene rings between them.

This is true of all cross-linked resins and is what makes these organic resins so useful in the composites industry. They have the same strength and stiffness in any direction. They are what are called isotropic materials—they have the same properties in all directions. This is quite different from the fibers that we talked about in the previous chapter, where the

Steps to making a cured orthophthalic polyester (drawing by the author).

strength and stiffness of the fiber is very high in the direction of the fiber and rather weak across the fiber. And when we get to design and analysis of composites we will cover this again. But I did want to introduce you to the concept of isotropic versus non-isotropic materials here so that later you will have an idea of what I mean.

What about isophthalic acid and isophthalic polyester resins? Well, we can't very well make phthalic anhydride out of this stuff—there's a carbon atom in the way. But, fortunately, the resin itself can be made using the same basic reactions as the orthophthalic polyester resins. It's just that you don't form phthalic anhydride initially. You react maleic anhydride with isophthalic acid and propylene glycol to form the basic backbone that can then be cross-linked with styrene. It looks something like the pic on the next page.

What results is a somewhat different structure than what we made with the ortho isomer of phthalic acid, but it still has the same basic properties: a long chain of esters with benzene rings attached to them and lots of reaction sites for cross-linking.

Isophthalic acid is more expensive than its ortho cousin, so this resin system is pricier than what you buy down at your local Home Depot. It does, however, have much better corrosion and chemical resistance, and

Synthesis of isophthalic polyester resin (drawing by the author).

it also doesn't uptake water the way ortho resins do. So, when you see a shiny fiberglass boat go by with its pretty white hull, the white that you see is usually an isophthalic polyester resin. This resin, used as what is called a gel coat, is sprayed into a boat hull mold until it is thick enough that it will be impervious to the water, and it is allowed to cure in the mold. Then an orthophthalic polyester resin is used to lay up the remainder of the fiberglass in manufacture of the hull of the boat. Anyone who has owned a boat knows that the inside of the hull (the bilge) looks very different than the pretty, shiny outside of the hull. They also know that you need to keep the bilge dry, or it will rot the fiberglass. This resin is also preferred over its ortho cousin for making the fiberglass tooling that a lot of other fiberglass parts are made with. It is dimensionally stable, and the added corrosion and chemical resistance makes it an ideal material to use for molds that are used over and over again to pump out fiberglass parts.

Vinyl Ester Resins

Now that we know all about esters and ester-making carboxyl groups, as well as how they can be formed from benzene ring-based organic acids, let's move on to the vinyl group. But first we need to know where this group comes from, so let's first look at ethylene. Ethylene is an organic compound with two carbon atoms (that's where the *ethyl-* comes from)

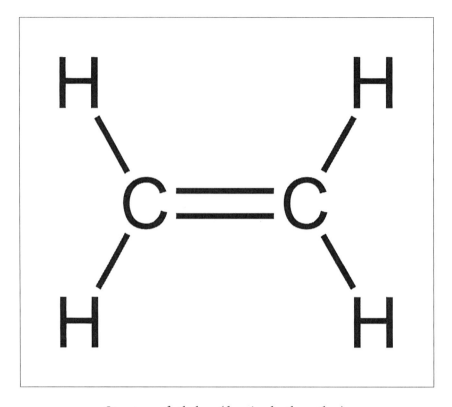

Structure of ethylene (drawing by the author).

with a double bond between the carbons. In its simplest form, this is what it looks like.

This odorless, colorless gas is the simplest of what are called olefins. Olefins are hydrocarbons wherein one or more of the carbon-carbon bonds is a double bond. They make up quite a few of the oils that we use every day in cooking, and when longer chain, more complex olefins become organic acids are called unsaturated fatty acids. We have double bonds again, this time carbon-carbon for olefins versus the carbon-oxygen double bonds in organic acids.

But back to ethylene itself. Ethylene is a very important molecule in the plant kingdom. Plants use it to signal leaves to turn brown in the fall and die off, to signal flowers to open up to be pollinated, and then to wither and have the petals fall off so that the seeds can form, and also to tell fruit to ripen. It is used in agriculture to control the ripening of fruits so that most of the fruit on the tree ripens at the same time—better to harvest your fruit crop at once than go to each tree to select the apple that's ripe that day.

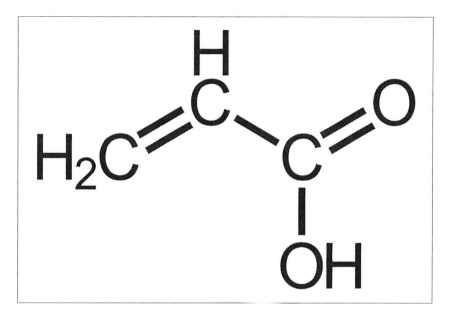

Acrylic acid (drawing by the author).

Vinyl esters are basically a vinyl group that has an ester-making carboxyl group attached to it. The simplest of these is what is known as acrylic acid. This acid is basically a vinyl group with an ester-making carboxyl group attached to the end of it. Of course, we learned above that the carboxyl group is what makes organic acids be proton donors, so this is definitely an organic acid. It looks like the following:

What is at the left end of this picture is a carbon attached to two more hydrogens to complete the molecule. This is the backbone of the plastics known as acrylics, or clear moldable plastics, the most common of which is called plexiglass.

Vinyl ester resins are made by reaction of these acrylic acids with an epoxy backbone like BPA diglycidyl ether to form a resin. Remember bisphenol-A? We learned about that when we were discussing the chemistry of carbon and why the structure of carbon compounds is so important to composites. I told you that we would get back to this when we covered the glue in composites, and here we are.

The vinyl groups in the resulting resins are very reactive because of the double bonds both between the carbons and on the carbon-oxygen bond, so they are prone to form together into polymer chains and become solid. To prevent this from happening, inhibitors are introduced in the manufacture of these resins to make sure that the product ends up as a liquid that is stable and won't spontaneously cure into a solid. It's easier

Typical BPA-based vinyl ester resin (drawing by the author).

to sell that way. In any case, the resulting vinyl ester resin looks like the above pic.

You will notice the bisphenol-A in the middle of this thing with a carboxyl group attached where the -OH group was (condensation reaction again), making it BPA diglycidyl ether (DGEBA). Then there's a vinyl group attached at each end of the carboxyl groups replacing the double bonded oxygen with the vinyl group. This is the most common vinyl ester resin precursor made. There will be more about DGEBA when we get to epoxies, since that is the most common backbone of epoxy resins as well.

The inhibited vinyl ester resin is then usually dissolved in styrene as a reactive solvent to make a liquid that can be used as a laminating resin for composites. This is where the epoxy resins and the polyester resins meet: with styrene. As we learned above, the polyester backbone is cross-linked with styrene which is again a compound with a benzene ring. In much the same fashion, vinyl ester resins are reacted with and cross-linked with styrene to form a solid. This time, instead of using a small amount of a hardener in the resin, the reaction proceeds through the introduction of what are called free radicals. These free radicals are basically just a little molecule that has one unpaired electron. The best example of this is the -OH or hydroxyl radical which is water without its other hydrogen. These free radicals are very short lived, so they start chemical reactions very quickly. In the case of vinyl esters, they can be made by irradiating the uncured resin with UV light, or by using peroxides like hydrogen peroxide, which is basically water with lots of the hydroxyl radicals floating around in it.

Vinyl esters, as you may be able to tell by now, are sort of a bridge between polyesters and epoxies, and they are used as such. They have much lower water absorption, are more chemically inert than polyesters, are stronger and stiffer than polyesters, and are somewhat more expensive than polyesters—even isophthalic polyesters. They are not as strong or stiff as epoxies, but they are less expensive. They also don't shrink when cured, so composite parts made from them are dimensionally stable. This makes the tooling much simpler for higher precision parts and structures. They are also stickier than polyesters, albeit not as sticky as epoxies. This is because they have better cross-linking and cross-bonding properties than polyesters, so they will chemically bond to other similar plastics rather than just sticking to them. This limits their propensity to delaminate, and they make a more permanent structure. They also have a lower viscosity than typical epoxy resins, so they are a more workable substitution for epoxies if the part being made allows for slightly lower strength and stiffness than epoxy. For all these reasons, vinyl ester resins are considered a bridge between polyesters on the lower cost, low-performance side, and epoxies on the higher cost, high-performance side.

Epoxy Resins

Now we are going to learn about another group, the epoxide group. So far in this chapter we have studied alkyl groups, carboxyl groups, ester groups, vinyl groups, and now we are going to look at epoxide groups. This is a very reactive group in which an oxygen is bonded to two carbon atoms in a triangular structure. Then other alkyl groups (remember, alkyl groups are those represented by *R* in organic chemistry) are bonded to this triangular structure. The basic structure of an epoxide (adapted from *Wikipedia*) is shown below.

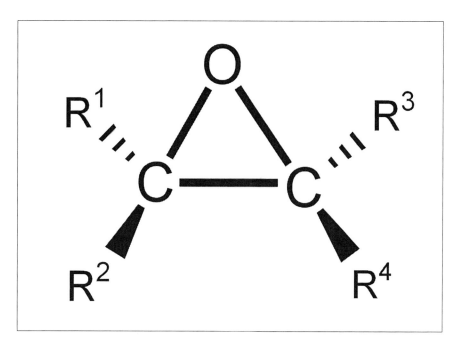

Basic structure of an epoxide group (drawing by the author).

And if you remember, when I touched bisphenol-A, I noted that I would get back to this very important compound when we got to epoxies. Well, BPA-based epoxies are the most prevalent epoxies on the market and dominate total global epoxy resin production. Bisphenol-A is not only the backbone for vinyl esters, but it is also the dominant precursor to most of the epoxy resins produced today. Bisphenol-A is a very important compound, which is why there is so much of it around. It has some rather deleterious health effects because it can imitate the body's hormones. It messes with the endocrine system in pretty significant ways, so there is a lot of public pressure to reduce human exposure to this molecule. But that's the subject for another book, so let's get on with epoxies.

To make an epoxy out of bisphenol-A, first we have to make it into an ether to get the oxygen atom that makes the epoxide group. This is done by reacting bisphenol-A with an organic chlorine compound (organochloride) called epichlorohydrin. The figure below shows the structure of epichlorohydrin in a stick figure form so that you can see the three carbon atoms, the oxygen, and the chlorine. Note the epoxide group here on the other side of the molecule from the chlorine. When these two are reacted together you get an epoxy resin and some free chlorine in the form of hydrochloric acid (HCl) if the reaction were allowed to go this way.

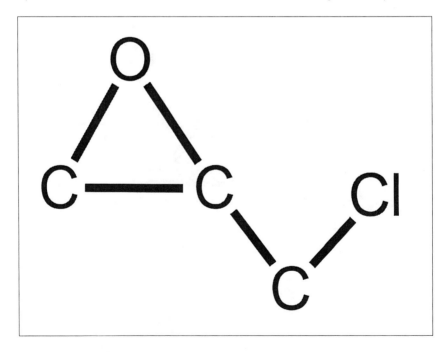

Structure of epichlorohydrin (drawing by the author).

In reality, the reaction is a bit more complex than this because there is an intermediate step that is introduced to make sure that it does not make hydrochloric acid. In the actual synthesis of the epoxy, once these two epichlorohydrin molecules are connected to the bisphenol-A, the chlorine is removed by performing this reaction in a water (aqueous) solution of sodium hydroxide. The resultant of this entire process is DGEBA and salty water. And remember, DGEBA is the backbone of vinyl ester resins as well, but vinyl ester resins don't keep the epoxide groups intact, whereas BPA-based epoxy resins do.

Our bisphenol-A is still there, only now it has two epoxide groups added, one on each end, and it's ready to connect up with more of this to make a solid plastic. And when this stuff is made, if you want a higher molecular weight resin—one that is a little stickier and less runny because

Formation of BPA diglycidyl ether (DGEBA) from BPA and epichlorohydrin (drawing by the author).

of how they want to make the part—all you need to do is to add more bisphenol-A to each end of this and you have it pre-polymerized.

In fact, you can add up to around 30 of these bisphenol-A molecules to this base diglycidyl ether, and you get a solid that can be ground into a powder. You can also play with the amount of epichlorohydrin you add to the mix to make sure that you get these epoxide groups at the end of the longer chains to make it more reactive. Or you can start with a liquid resin and add more bisphenol-A plus a hardener and semi-cure it at about 320°F to get an even higher molecular weight epoxy.

What happens when you do this is you reduce the epoxide group count on the epoxy resin, so it starts to behave more like a thermoplastic and less like a typical epoxy. Since the resins still possess the -OH groups all along the backbone, they can be cross-linked to produce a solid that has the properties of a typical cured epoxy resin—hard and rather stiff. They just don't cure as easily or as quickly as shorter chain epoxies. As you can see, epoxy resins come in many variations on a theme which makes epoxies very useful for solving any number of different composite design problems.

Other Epoxies

Another important class of epoxies are the Novolac-based epoxies. Remember phenolic resins? Well, the same chemistry that makes Novolac phenolic resins can be used to make epoxies by just adding an epoxide group onto the benzene ring. And we can use the same chemistry we used with bisphenol-A—as in reacting with epichlorohydrin—to produce the epoxide groups that make this an epoxy resin. Becayse there is only one benzene ring in a Novolac, whereas there are two in bisphenol-A, Novolac-based epoxies usually have lower molecular weight and are used as adhesives, to make tires stickier, etc.

Since this resin is phenol-based rather than BPA-based, it is less expensive. It is intermediate in price between the higher end BPA-based epoxies and vinyl ester resins. In fact, the West System epoxy that every boat owner has used at one time or another is a Novolac-based epoxy resin. It is advertised as a replacement resin for polyester or vinyl ester resins, especially for making repairs to your boat. You can even thicken the resin and use it as a fairing compound to fill in dents and such, or make it even thicker and use it as an adhesive for gluing stuff together on your boat.

There are other types of epoxy resins, ones made from aliphatic hydrocarbons (hydrocarbons without a benzene ring) with epoxide groups attached to them so that they polymerize. These aliphatic hydrocarbons can have ring structures, but the rings are bonded together rather than being separated into benzene rings or phenol groups. These resins do not react as easily as the phenol-based or BPA-based epoxies because they lack many of the -OH groups that the phenol or aromatic epoxies do.

One class of these aliphatic epoxies is made using vegetable oil as its backbone—typically soybean oil. These resins are cheap to make, so they are used often as diluents in other plastic products that don't need the strength and stiffness of the aromatic-based epoxy resins. One common use is as a secondary plasticizer and cost stabilizer for PVC, the stuff of sprinkler pipes.[6]

Epoxy Curing Agents

But how do we make a solid out of these liquid resins? Anyone who has messed with epoxies knows that they are almost all a two-part system: the base resin and a hardener. The most common hardeners are amine hardeners.[7] Now we need to learn about what an amine or amine group is.

So what is it with groups in this book anyway? Well, organic chemistry is all about what common groupings of atoms are seen over and over again. They become the language of organic chemistry because each group performs a similar function when attached to a larger molecule, and each group is also the site of where the majority of the chemical reaction magic happens. Control over these groups and understanding them is critical to understanding organic chemistry. And since all composite glues are organic molecules, I talk a lot about groups.

On with a description of amines and what they do. We learned a little about amines when we touched on nitrogen in the chapter on the periodic table, so it should come easily to you that amines are compounds with an amine group. This group is a nitrogen atom with two hydrogens attached to it, which can then be attached to almost anything else. Common household ammonia is the simplest of these compounds because it is basically hydrogen amine. And most amines used in industrial chemistry are derivatives of ammonia, so that's why I used ammonia as my first example of an amine. Primary amines have two hydrogen atoms attached to a nitrogen atom, and the third place on the nitrogen atom is taken up by something else. Secondary amines occur where one of the two hydrogens is replaced with something else, and finally tertiary amines are where all the hydrogens are replaced with something else and there is just a nitrogen atom in the middle.

I went through this explanation of amines first so that you would understand how epoxy resins can be hardened using amines. If you introduce a primary amine to an epoxide group, the primary amine it attaches to that group as a secondary amine and one of the hydrogens that was

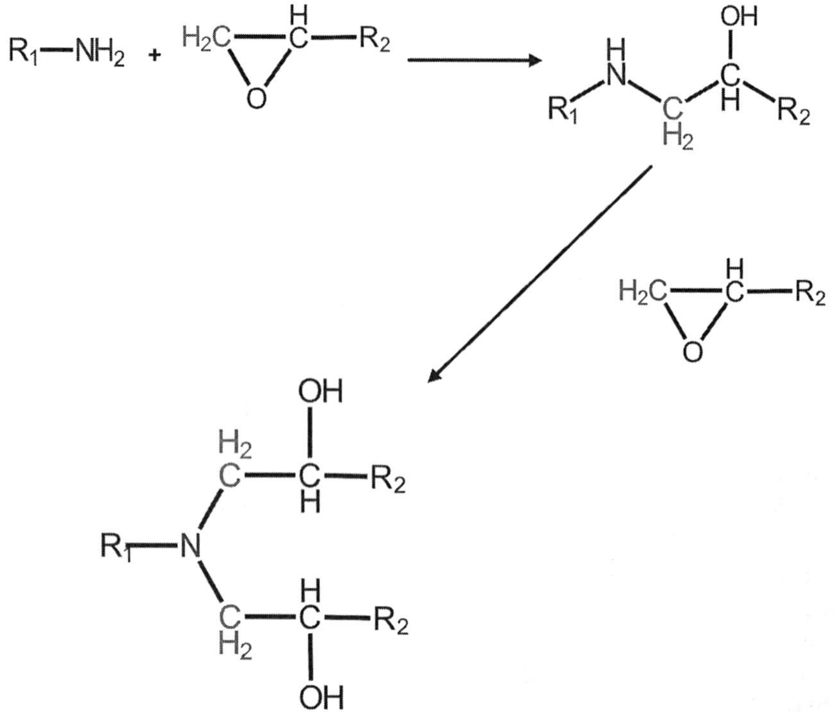

Amine curing of an epoxy resin (drawing by the author).

attached to the nitrogen atom attaches to the oxygen in the epoxide group. Then you end up with a hydroxyl group where you had an epoxide group. Then you can do this again, forming a tertiary amine where the nitrogen sits in the middle between three hydrocarbon groups. It looks like the following:

The great thing about all these Rs is that they can be any aromatic or aliphatic hydrocarbon. And if you use a double-ended epoxy like DGEBA, you get a three-dimensional matrix of benzene rings all tied together to make a solid mass. And this solid mass is resistant to most chemicals, impervious to water, hard but not brittle, and sticks really well to any surface. This is the magic of these materials and why they are used so heavily.

Amine-based curing or hardening agents are the most common curing agents used in epoxy resins in industry today. The most common of these are called diamines where there is a hydrocarbon group in the middle—usually a short chain hydrocarbon—with an amine group at either end. When you react DGEBA, with its epoxide groups on each end, with one of these diamines, a molecule of diamine attaches itself to each end of the DGEBA. And you now have the beginnings of a polymer, because more

Start of the hardening or curing reaction of a BPA-based epoxy using a diamine curing agent (drawing by the author).

DGEBA can attach itself now to the six hydrogens that are available on the amine groups to attach to. The hardening reaction starts like this:

The cross-linking to make a three-dimensional plastic happens at each end of the DGEBA where you see the hydrogens sticking down in the picture above. Since there is lots of DGEBA around in comparison to the amount of diamine hardener, the excess DGEBA attaches to these hydrogens and can attach to another chain of the same stuff—cross-linking at its best.

But, of course, there are other ways to cure epoxies so there are several other types of hardeners to use depending on what you want the final epoxy to do or what properties you're looking for. You can use anhydrides (remember maleic anhydride?) to cure epoxies thermally at high temperatures. These are good candidates for making electrical insulators where you have to incorporate a lot of mineral filler in the epoxy and you want it to take a long time to cure.

You can also use phenols and polyphenols, like Novolac and bisphenol-A itself. This has to be done at high temperature and in the presence of a catalyst to make the reaction work, but the resulting product has higher thermal and chemical resistance than epoxies cured with amines or anhydrides. The Novolac-based hardening agents come in a solid powder form and are used quite a bit in powder coating, which is why powder coated parts are so resistant to corrosion, scratching damage, etc. This stuff is very hard and nearly impervious to chemicals, heat, water, you name it.

And, finally, you can also cure with thiols or sulfur-bearing organic compounds. The sulfur reacts very easily at room temperature with the epoxide group, so these are typically two-part epoxies cured at room temperature. They don't have nearly the chemical or thermal resistance of other epoxies that are cured at a high temperature, but they cure very quickly. They are used for quick-curing epoxies like 5-Minute Epoxy. And if you've ever used this product, you can smell the sulfur as the thiol-cured quick-setting epoxy cures. A lot of this stuff is used in the construction industry in California for gluing bolts to existing foundations for earthquake protection in homes.

Thermoplastic Resins

Finally, we come to the thermoplastics as composite resins. These plastics are a recent addition to the mix, but they offer a few advantages—mostly in automated processing and fabrication. By and large, thermoplastics are made into a pre-impregnated product (prepreg) where the fibers and the plastic matrix are bound together by the prepreg supplier.

There are a number of thermoplastics that are used as matrix materials—or glue—for composites. Some very common ones are polyethylene and polypropylene. These two thermoplastics also make good fiber if the molecular weight of the polymer is high—as in ultra-high molecular weight polyethylene—a.k.a. Spectra and Dyneema fiber. If you remember, I covered quite a bit of information about these two compounds and their fibers in the chapter on strings.

Some other thermoplastics that have quite a few applications in composites are polybutylene terephthalate (PBT), polyamide (PA), polyphenylene sulfide (PPS), and polyether imide (PEI). Diagrams of the structure of these four are below.

Structure of some other thermoplastics used as composite resins (drawing by the author).

Note that all these thermoplastics have embedded in them that 6-carbon benzene ring structure that we love so much. This is what makes them strong and tough materials that are very useful in the composites industry. Each of these has different processing requirements, meaning melt temperature, viscosity of the melt, stickiness of the molten or softened thermoplastic—all the things that are important to automated fabrication processes. Most of these thermoplastics can be extruded molten and rolled into sheets the right thickness for laminating into a layered composite. But we will get to that when we get to fabrication processes, or how to make something out of composites.

Two more thermoplastics that need to be touched on a little bit are relatively new to the marketplace—at least compared to most other plastics. They both have weird names as well. They are both ketones, which is another very important functional group in organic chemistry. A ketone is basically two hydrocarbon groups—one on each side of a carbon with a double bonded oxygen attached to it. Like esters that have a hydroxyl group (-OH), ketones have a double-bonded oxygen. Acetone—the basic solvent in lacquer thinner—is the most common ketone that most of you are familiar with.

These two plastics with the funny names are PEEK and PEKK, which if you say them quickly sound sort of funny. Their actual names are polyetheretherketone and polyetherketoneketone. They look like the following:

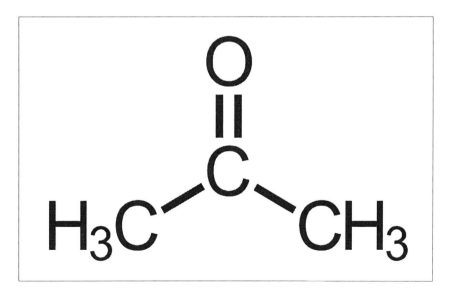

Structure of acetone (drawing by the author).

Structure of PEEK and PEKK (drawing by the author).

See how similar they are: PEKK has one double-bonded oxygen atom and PEEK has two. PEEK was developed in 1978 and was introduced on the market in the early 1980s as Victrex by Imperial Chemical Industries.[8] It has since found its way into some very high-performance composites. PEEK is also bio-compatible, so it is used in some bone and joint replacements because of its high strength, resistance to abrasion, and self-lubricity—it is a great material for bearing surfaces. It is, unfortunately, rather expensive compared to other plastics, so it is only used in very demanding applications. Both plastics have high heat tolerance and are classed as high-performance thermoplastics. This is mostly due to the nature of their structure; they are mostly composed of a series of 6-carbon benzene rings strung together with either a carbon double-bonded oxygen, or just an oxygen atom connecting the polymer together.

And that brings us to the end of this chapter on glue. Hopefully you learned something, and you can always refer back to this when you get to the part of the book where I discuss design and analysis of composites.

We need to move on to the third major branch of the semantic tree of knowledge of composites: how to make a composite with our newly developed knowledge of the strings and the glues that bind them together.

5

Making Something Out
of Composites

Now that we know the basics of where the fibers (string) and resin systems (glue) come from, how each is made, and also some of the idiosyncrasies of both the strings and the glues, we need to know what to do with the string and glue to make something. In other words, we have to address the last major branch of our semantic tree of knowledge about composites so that our tree is balanced. Once we have all three of the major branches—the string, the glue, and what to do with them to make a composite—we finally have the basics of the semantic tree of knowledge for composites and can start to build out branches, leaves, and even flowers and fruit.

But before we can understand how to blend the strings and glues together to make a composite, we have to know what product forms we can get from our suppliers. If you haven't noticed, there is a very practical side to this book because the author is a practical sort of person who likes to get down to the basics and go build composites.

String and Glue Product Forms

Just like the infinite number of potential combinations of different strings and glues, there is a nearly infinite array of different starting product forms available to the composites designer. Fortunately, they break down along some predictable lines. This is because the string component is essentially a textile, and the glue part is a plastic. These can be further broken down into thermosets and thermoplastics, like we noted in the glue chapter of this book. For the glues, thermoset resins come in liquid or dry powdered form, and are either a one-part plastic precursor or a two-part plastic precursor where the resin and hardener need to be mixed immediately prior to using them. Phenolics commonly come as a one-part resin system and are powders that are pre-mixed with exactly the right amount of hardener. These powders are kept cold to make sure that they stay in powder form and don't react to become a hard plastic before they can be used in a composite. Polyesters, vinyl esters, and epoxies more typically come in liquid form and are two-part systems that are mixed just prior to using them. The 5-Minute Epoxy you can buy at Home Depot and the West System epoxy created by the Gougeon Brothers, Inc. in Michigan are good examples of this product form. The strings come in continuous filament as tows for graphite and most organic fibers, and strands for continuous glass fiber. The strings can be woven into fabrics of all sorts, thin or thick, with more fibers in one direction than the other, etc. Strings can also be chopped into short lengths and either pressed into a mat or mixed with resin and sprayed into a mold.

Let's focus on glass strings first, since they are the most heavily used fibers in composites today. Glass fibers come in different weight strands, meaning there are more or fewer fibers in each strand. There are some glass yarns which have twisted fibers that are not typically used for higher performance glass composites because the twisting of the fiber reduces strength and stiffness of the laminate. Thicker, untwisted yarns are called *rovings,* which are untwisted bundles of continuous fiber. Untwisted yarns (I'm just going to call them yarns from here on out) and rovings have different weights based on the filament diameter as well as how many filaments make up the strand or roving. Weight of these product forms is described in units of *tex,* which is the weight in grams of a 1000 meter length of the yarn or roving. Yarns typically have weights in the range of 5 to 400 tex, whereas rovings have weights in the range of 300–4800. Rovings are typically heavier than yarns and are used primarily to build up unidirectional or woven layers of glass more quickly than a unidirectional or woven yarn. And what's with this weaving stuff? Yes, glass fibers come in a number of different fabric forms, from a completely balanced weave where there are the same number of glass fibers in each direction in the cloth to a nearly unidirectional product where heavy weight rovings are woven with very small rovings going the other direction—just enough to hold the cloth together. And, of course, there is everything in between. There are satin fabrics where the cross-weave rovings (weft) skip every other one of the in-line or *warp* rovings. And there are also twill fabrics that have yet another weave pattern. And these woven rovings come in different weights and thicknesses based on where they are intended to be used. Some glass cloth can be rather heavy since it is made of the highest fiber count rovings that the weavers can handle. The weight of the glass cloth is sometimes reported as yardage count, which is how many yards of the stuff it takes to weigh one pound. Thicker glass cloth with heavier rovings has a lower number of yards per pound, so it has a lower yardage count. This might seem counterintuitive at first, but that is the way it is sold, at least internationally. Here in the United States, you can buy fiberglass cloth by its actual weight per yard—as in 4 oz. cloth. And you can get up to 24 oz. glass cloth on Amazon. But we are not weavers here, so all we need to know is that there are these different weaving patterns and different weights of cloth available to us.

In addition to continuous filament products, especially for glass, there are chopped fiber product forms as well. A typical chopped fiber mat is made up of 1–2" long chopped fibers dropped onto a moving belt, then sprayed with a small amount of a binder and rolled into a flat sheet. Just the random orientation of all the tangled fibers, along with the small amount of binder that is added, makes for a sheet that can be easily cut to the desired shape and draped into a mold. This is the material that is used to make the surface of a surfboard so smooth. It is typically the first layer applied over the gel coat in a boat hull mold to make sure that the gel coat has a strong and very smooth backing before more of the thicker-woven roving is applied to build up the boat hull. But I get ahead of myself here.

Another very important product form for not only glass but for all composites is what is called unidirectional tape. This is a fiber form that comes in a flat sheet of continuous fiber all aligned in one direction and held together with just enough binder that it keeps its shape on a roll of the tape. These tapes can be as narrow as an inch or so and as wide as two feet. The wider tapes are typically held together with a woven weft thread every inch, or a very thin bead of hot melt glue laid across the tape at regular intervals to keep all the fibers aligned, in addition to whatever binder is used in manufacture to allow the finished tape to be handled easily. Narrower tapes are usually held together with just a small amount of a binder to keep the fibers aligned until they are ready to be used.

The glues, as we have already learned, come in liquid or powdered form, and some thermoplastics come in sheet or film form, where the sheet or film is intended to be used between two layers of either fiber cloth or unidirectional tape. Typical polyester, vinyl ester, and epoxy resins are sold in liquid form where the fabricator is required to mix a pot of the resin with its hardener. This brings up what is called pot life, which is a crucial input to any fabrication process.

Resin pot life is determined by many factors. First is the resin itself and what its polymerization (hardening) chemistry is. This determines how the polymerization cross-linking reaction takes place and whether it is a more gradual process or happens very quickly once it starts. Second is what temperature the resin and hardener need to be to start the reaction and have it go to completion. In general, once a reaction starts, higher temperatures make it happen quicker.

Of the liquid resins that need a catalyst or hardener to initiate the reaction, there are room temperature-setting resins and higher temperature-setting resins. In general, the polyester and vinyl ester resins and some epoxies harden at room temperature. These resins are used for things like boat hulls and large wind turbine blades where it is difficult or impossible to put the entire structure into an oven. The higher performance BPA-based epoxies are almost all higher temperature-curing resins, with some epoxies curing best at 300 to 350°F. These are advanced resins and are used mostly in the aerospace business and for high-performance structures, like fighter aircraft wing skins. Most of these parts are cured in an autoclave where the part is put into a plastic vacuum bag, and heat and pressure are used to squeeze out as much of the excess resin as possible and consolidate the part.

There are also solid resins that are ground into a powder. Then a hardener is added, and they are applied in many ways. One way is to use a powder printer, much like applying a powder coating to your favorite metal part. There are some epoxy formulations that are mixes of the powdered resin and powdered hardener. Mostly, however, it is Novolac phenolic resins that come in powder form and that are hardened by the addition of the acid catalyst. If the catalyst is already mixed with the Novolac, these powders need to be stored at very low temperatures because they will react and harden over time if kept at room temperature. There is, therefore, somewhat of a storage life issue with some of these resins, and this needs to be taken into account when using them to make a composite part.

Thermoplastic resins come in either powder form or film. The powders are typically used for chopped fiber composites and are mostly mixed with chopped glass fibers. The films can be used to make a laminate by putting down a layer of unidirectional fiber, then a sheet of thermoplastic, and then another layer of fiber and so on until the thickness is built up to what you want. Then the stack is heated and compressed between two sides of the mold, and the thermoplastic melts and gets infused into the spaces between the fibers to make a solid. Then when the laminate is cooled to room temperature, or a temperature below the temperature where the thermoplastic softens, the part is taken out of the mold as one solid piece.

One more product form that is typically used in advanced composites in the aerospace industry is a prepreg sheet of fiber and resin where the resin is not completely hardened. This is most typical with carbon-epoxy composites where the epoxy is taken to what is called the *B-stage* cure, where it is gelled up a little bit but has not completely hardened. These sheets can then be cut to shape, put in a mold, stacked up to make the laminate and then cured in either an oven or autoclave to make the laminate.

The wing skins of the F-35 fighter aircraft are made this way, using a pre-preg of carbon fiber and a high-performance epoxy resin.

And there is a myriad of choices when it comes to prepregs. There are glass fiber prepregs with an epoxy in its B-stage, but mostly the market for prepregs is carbon fiber. They come in sheets of unidirectional fiber, balanced and non-balanced fabric, and every combination and permutation of fiber/matrix combination that has been found to be useful to someone. There are even prepregs made from thermoplastic matrices where a thin sheet of a thermoplastic—like PEEK, for instance—is placed on a tape or fabric of carbon and heated between two platens until the resin melts into the fibers and completely wets them. These are pretty stiff so they are not as commonly used as epoxy prepregs, but they do offer a processing advantage if you can use them. Since they are thermoplastics, the resin does not need to cure to make the part, all you have to do is put some sheets of the thermoplastic matrix prepreg in your mold, close it up, heat it and apply pressure, and presto—a composite part. Boeing[1] has been working with thermoplastics and thermoplastic prepregs for some time and has studied them extensively for use in aircraft and spacecraft.

Before we go on here, I need to address my interchanging of "resin" with "matrix" in discussions of composites. I do this because I come from a design and analysis background, wherein when talking about mechanical properties of these wonderful materials, we often cite the fiber properties versus the matrix properties. And we get tired of using the term "cured resins" all the time, so the common vernacular is to call the cured resin that is incorporated into the composite the matrix. People who are fluent in the analysis of composites quite often talk about separation of the fiber-dominated mechanical properties from the matrix-dominated mechanical properties. I elaborate on this when we get to that part of the book, but I wanted to make sure that nobody was confused by my calling the cured resin the composite matrix.

Composite Molding Processes

Molding is the most common way of making a composite. In this process, the starting materials—the strings and the glues—are placed into a mold cavity either by hand or by an automated laying machine, and the whole part is cured in one piece. Each molding process is quite a bit more complex than this, but this is the essential process, so you get the drift here. We need to cover all the different molding processes—from the oldest and most manually intense to the newest and most automated—and at least touch on everything in between. The intent here is not to make you experts in all these processes—that would require an entire bookshelf. Rather, what I plan to do in the next several pages is describe the most common molding processes from the point of view of each being an extension or its own modification of the simple processes laid out at the start of this paragraph, and tying them all together so they make sense. Let's get to it.

Hand Layup and Open Molding

This is the traditional method of making something out of composites and has been used in the recreational boat building industry since the dawn of fiberglass boats. Most people are aware of this way of making boats, but since this process is really the foundation for all molding processes, it is important to understand it and how all the other composites molding processes spring from it. Hand layup is the most labor-intensive process for making a composite because each layer must be placed by hand.

Hand layup is what is called an *open mold* process because the mold is one-sided and open at the top where the fiberglass is being laid down

Example of a boat hull being extracted from its mold (reproduced by permission of Isaac Oczkewicz, La Conner Maritime Service).

inside the mold. This is still one of the most commonly used methods for making most of the parts for a boat, where the hull is laid down in one mold, and the top sides or deck are laid down in another mold, and the two halves are bonded together. It goes without saying that all the bulkheads, stiffeners, etc., are also bonded into the boat hull while it is being laid up. Very large sailboats and powerboats are still made using this method. With the assistance of cranes and other material-handling equipment, this process can be easily expanded in size to make some pretty large things out of your composite.

Hand layup is still used for a number of reasons. It allows you to place the fibers very precisely and in a very complex layup. It also allows for the addition of a core which is very important not only for buoyancy, but also to make the structure stiffer, which is good for a boat deck. These core materials traditionally were balsa or sometimes plywood in heavier boats, but today boat builders have moved to PVC foam so that their core has no chance of rotting.

The closest variation to the open mold process is what is called *vacuum bagging*. In this process, the resin is not allowed to harden completely until the entire part has been laid up. Then a plastic sheet with holes in it for vacuum lines is placed over the top of the laid-up part in the mold. Then a vacuum pump sucks all the air out of the bag and effectively closes up the mold. This applies pressure to the uncured composite and consolidates it, eliminating all the entrained air bubbles.

If a room temperature resin, like polyester or vinyl ester, is chosen for the matrix material in the open mold or open mold plus vacuum bag process, there is almost no limitation on the size of the part that can be made. The only limitation on size is how large a mold you can make and how quickly you can lay down the glass and resin. In Denmark and other Nordic countries, they have built fiberglass ships for their navies that measure up to nearly 200 feet long. These are mostly sandwich constructions, but the hull is made all in one go in one mold.

These open mold processes always need to be done in a clean environment, and the new layers of glass and resin need to be laid down on relatively fresh and not completely cured material. Resin adhesion is very much a surface phenomenon, and the surface of the part that you are adding material to needs to be sticky for good adhesion.

If you are making a smaller part (a 30-foot boat hull, for example) this usually is not much of an issue. However, for larger boats, or for an entire aircraft wing for instance, the pot life and state of cure for the resin becomes very important. This is because if the layer you're about to add material to is completely cured, there is a chance that when you add more glass and resin (or carbon and epoxy) to thicken the part (boat hull for example), the composite may delaminate at the site of the added layers of glass and resin. If you remember cross-linking from when we covered glues, you can see that you need some uncured resin at the surface where you are adding more material to get the cross-linking to work so you don't get a delamination.

In the old days, there was little to no personal protection, and the workers who made fiberglass boats were exposed to the resin fumes daily. These fumes are primarily caused by the styrene backbone and the organic peroxides that are given off by the polyester resins—pretty nasty stuff. The boat building industry came under scrutiny by the EPA in the late 1990s, and by 2000, the EPA, with input from the boat building industry, proposed a ruling that would eventually control the toxic emissions that are given off by the polyester resins that they used.

Nowadays, all boat production has gone inside, and is done under tightly controlled conditions. These factory floors are kept clean and free of all the glass and resin dust, as well as the detritus pieces and parts that used to be strewn about. Employees in these factories all wear special powered air filters so that they are no longer exposed to the resin fumes. While some spillage of resin does happen, that spillage is a cost to the boat builder, so it is minimized as much as possible.

All this is also true of the modern composites manufacturing and fabrication business. The environments in which all composite parts are made are tightly controlled, and the people who provide the hand labor are very skilled technicians who understand both the structure and the chemistry of what they are doing to make these composite parts.

What are some other examples of molding processes? The open mold, vacuum bag process is used extensively in all composite fabrication. For higher end structures like wind turbine blades, aircraft wings, and flight control surfaces, the layups of fiber and resin are very tightly controlled and there are quality checks all along the way. But we will get to composites design and why certain layups are used when we get to the composites design chapter later in the book.

There are many variations on the hand layup process, and it is still ubiquitous in the composites fabrication industry. It is, in fact, the standard method for making most boats, aircraft parts, etc. There are also several variations of this process where hand layup is used for part of the process of making a part (boat hull, for example), then to build up thickness where that is required, a spray-up process using chopped fiber and resin is used. What gets sprayed on the part to build up thickness can be a foam material or foam mixed with chopped fiber and resin to make a cored structure. Then, once the thickness has been built up to what the part requires, more layers of fiber and resin are used to make the interior surface of the part.

To make a lighter weight and stronger part, such as for an aircraft wing skin, the molds are somewhat more complex. These parts are usually made from carbon fiber and epoxy, and are usually laid up using prepreg, either in unidirectional form or fabric form. The molds for these parts must be able to withstand the pressure and temperature of the autoclave without warping because what is called the *outer mold line* of these parts is held to tolerances on the order of a few thousandths of an inch. This is critical for the aircraft industry because the shape of the wing must be exactly as the aircraft designer wanted it or the performance will suffer. It is also true of the auto industry, where fit and finish of body panels is critical. Auto customers demand that their cars look pretty, so the body panels and their design, fabrication, and finish are very tightly controlled. This also makes them too expensive for a hand layup process, which is why the auto industry has spent so much money in recent years to automate the manufacture of parts using lighter weight composites.

These higher end molds for higher volume production parts are typically made of metals and are called hard tooling, rather than the normal tooling associated with lower volume production or one-off parts. The low volume or one-off sorts of molds are typically crafted using composite materials in much the same manner as the part that is made using them. They are usually stiffened with kiln dried lumber to ensure that the dimensions will remain stable over time. Wet lumber shrinks, after all, and it tends to warp as it dries, which will cause all sorts of problems with the mold.

Most of the original hand layup parts—boat hulls, etc.—were made using what is called *wet layup*. In this process, the fibers in their fabric or unidirectional tape form are laid in the mold and the resin is brushed, rolled, or sprayed in to wet the fibers completely. Then the next layer is added to that wet layer and rolled in with wet resin take care of any dry spots. This process is repeated until the thickness of the laminate matches what the part designer wanted. Sometimes, as mentioned above, to make the process go more quickly, once the first few layers are added and allowed to gel, chopped fiber and resin is sprayed into the mold to build up thickness. This is only used when the structural needs of the part don't require a continuous fiber laminate. But with this sprayed up material, small pieces of foam can be added to the spray mix to create some additional buoyancy or a lighter part if needed. This, in addition to using foam core, can create a light weight, stiff structure using an inexpensive and less labor-intensive process.

The process for making the mold is something like a cross between an exact science and an art form. The ideas and some of the techniques were born in the metal stamping industry for things like auto body panels.

These metal stamping molds are very expensive, but they are made for use up to a million times, so it's worth the cost to get them right. For composites, on the other hand, they aren't used more than a few to maybe a couple of hundred times at most, so molds for composites are typically made of less expensive materials and lighter weight, easier to form materials like balsa and rigid foam. Foam is nice because it can be easily shaped, is cheap, dimensionally stable, and you can glue together as many blocks of it as you need to make a mold for your part.

The process for making a mold starts with making what is called a *plug*. This plug is an exact representation of the shape of the surface of the part you want to make. This is the part that is typically made out of the foam. Your plug must be an exact replica of what you are trying to make and should have a better surface finish than what you expect to produce in the mold. This is where the art in mold making comes in. There are artisans in the composites industry who can perfectly make these plugs. It takes quite a bit of skill to make a larger plug, but fortunately the careful home hobbyist who has good wood working and shaping skills can usually achieve a pretty good result with something the size of a day sailor, canoe, rowboat, etc. Small fiberglass aircraft have been built in people's garages, so it's entirely possible and well within the skills of a home builder as long as you are careful.

Making the plug takes some time because the surface has to be perfect. It must be better than the surface finish that you are looking for in your composite. This is because that surface gets transferred to the mold which then gets transferred to the part you are trying to make. Any imperfections in the surface of the plug will show up in the final part.

Once you are satisfied with the plug, now you have to make the make the mold itself. Making the mold starts off basically the same way you are going to make the part—with a few extra steps involved. First, you have to spray on a gel coat so that you have a smooth interior surface for the mold. This gel coat must be quite a bit thicker than the gel coat on your finished part needs to be, because you are probably going to have to do some touch up sanding on the mold surface itself before you can use it. Next, you need to lay down several layers of fiber and resin—fabric is the most commonly used material here because you aren't so much seeking light weight as you are structural rigidity and for the mold to be dimensionally stable. Then you need to bond in the structure that you are going to use to support the mold when you turn it over and set it on the floor to make your part. This is commonly done with lumber bonded to the outside of the mold so that you can make a rigid structure that will sit flat on the floor or will mount on whatever stand or tooling you use to support the mold when you are making the part. A lot of planning ahead is needed to make sure that you have the right support and that your mold won't warp, bend, or move around on you as you are making the part.

Now that we have our mold made, it is removed from the plug, turned over, and touched up. Here is where another piece of the art form of mold making really is important. The interior surface of your mold has to be exactly what you want the exterior surface of your part to be—including all the fit, finish, and other details needed. And since your eventual part is probably going to attach to some other part of the structure, the interface for your part where it attaches has to be built into the mold.

It goes without saying that when you are done making the part you have to get it out of the mold. Here is where the complexity of some molded composite parts comes in to play. Typical open molds that make shapes like boat hulls or wing skins don't have much of a problem because they can have what is called *draft* in the mold. Draft means that the opening at the top of the mold must be larger than the bottom of the mold and it needs to get progressively larger in all dimensions as you go up the sides of the mold. This will allow you to get your part out of the mold.

But what if you want a compound curvature for your part? Well, your mold just got complex, because you have to have at least a two-part mold to be able to do this. These molds are built by professional mold artisans. What they do is identify what are called *parting lines* in the mold. This is where they cut the mold very carefully and reinforce and finish the edges in such a way that they can bolt the mold together and it will fit as if there was no seam along the parting lines. Then, to make the part, the mold is bolted together and checked very carefully for inside finish, and the part is laid up in the mold. Then, when the part is fully cured, the mold is unbolted and the top piece(s) of the mold is removed and the part is carefully taken out of the rest of the mold. Very complex shapes can be made out of composites using multi-part molds, but these molds are expensive and time consuming to make, so people typically try to shy away from too complex a shape and figure out how to get their shape another way—like making it out of several parts and seaming them all together after they are fully cured.

Closed Molding Processes

Enough about hand layup and open molding. There are other molding processes to examine here, and variations on this theme are nearly infinite. If you want to have a decent finish on both sides of a part, and you want to make lots of them and make them more rapidly, a closed mold process is definitely one way to go. These processes run the gamut, but basically the idea is a two-part mold where the part is laid up in one half of the mold and then the other half is brought down against the uncured top of the layup. Then the two halves of the mold are pressed together to consolidate the part before it is cured. This is typically called either closed molding or compression molding depending on how much pressure is applied to consolidate the part. This process can be used with any reinforced plastic, including thermoplastics.

Closed molding is commonly used when there is a need to make lots of parts that have to be finished on both sides. It is also used in fiber reinforced thermoplastics to make the shape of the part. Here you need to add both heat and pressure to the mold to make the thermoplastic take the shape of the mold. These molds need to be made out of metal. And fortunately, metal molds have been used in the metal stamping industry now for more than a century, so the techniques for making these molds are well within the capabilities of most sheet metal mold makers.

Typically, closed mold processes are used for making either very large parts, like commercial aircraft wing skins and fuselages (think Boeing 787) and large wind turbine blades, or in automated manufacturing processes where thousands of parts are made using the same tooling. The mold itself is called a tool, and closed molding uses what is called hard tooling—as in metal tooling. Hard tooling can also be used in open molding, but because of the cost of making hard tooling, it is usually not cost effective to make a metal mold even for large parts if you don't want to make thousands of them. And there is also the problem of thermal expansion differences between carbon fiber composites and metal tools. Since carbon fiber composites usually involve an epoxy resin that needs temperatures on the order of 350°F to cure, the differences in thermal expansion between the part being cured and the tool that it is cured on make holding tolerances on the composite part very difficult. Especially because carbon fiber shrinks when you heat it, and metals normally expand when heated. There is a metal, a nickel-iron alloy called Invar, that has thermal expansion properties that are a good enough match with carbon fiber composites. But, because Invar is expensive and very difficult to machine, Invar tools are the costliest hard tools that you can make,[2] so these tools are typically only used for very high-volume production parts. These high-volume production parts also use automated processes for making the layup itself. But we will get to automation a little later.

Also, with the advent of 3D printing of plastics and even metals, several composites manufacturers have moved to 3D printing at least parts of their molds. And, if you have a large enough 3D printer that can print a material that you want to make your mold out of, you can 3D print the entire thing. There is one 3D printer made in Sweden by BLB Industries called THE BOX Large[3] that will print a plastic part as large as 2 meters by 2 meters by 1.5 meters. This printer uses plastic pellets and fuses them together as it lays them down to make the desired shape. There is one company, called Sciaky, Inc.,[4] that uses electron beam deposition of metals to make parts as long as 19 feet in their largest machine. Sciaky, based in Chicago, has been in the welding business since 1939 and pioneered much of the electron beam metal deposition technology that is commonly used today for high-performance aerospace metal parts. And, since this process can use nearly any metal and can form the alloy while it is printing, it is becoming more widely used in today's high-performance composite tooling business. Because this stuff is expensive, it is pretty much left to the Department of Defense and NASA, along with their contractors. That is, big aerospace—Lockheed, Northrop, Boeing, etc.

Now that we have our two-part closed mold, what do we do with it? Again, this opens up an entire range of possibilities. Initially, the closed mold process was used to make hand laid up parts where the finish had to be nice on both sides of the part. Then folks started thinking about how to automate the manual part of this, the hand layup process, and some automated tape laying machines were built that would lay up sheets of pre preg in the bottom part of the mold. Then workers would bring the top half of the mold to what was laid up and bolt the two halves together. Usually this involved compressing the part somewhat to consolidate it and to push out the excess resin. This process was fine tuned to the point where the composites designer could specify exactly the fiber/resin ratio in the part and the process could make it to those exacting specifications. A lot has changed in molding since the early days of hand building a fiberglass boat outside in a makeshift mold using just hand labor and brushes and rollers for the resin.

To make much larger parts out of resins that don't cure at room temperature, once you have the part or even large assembly laid up and the mold closed, you have to cure the resin. For aerospace parts this is usually done in an autoclave, which is basically just a really big pressure cooker. Of course, the pressure and the temperature are very tightly controlled, but it is still just a really large pressure cooker. One of the best examples of how advanced this process can get is the fuselage of the Boeing 787 Dreamliner. The fuselage of that airplane is one piece, and it is cured in the largest autoclave in the United States. The autoclave is at Spirit AeroSystems in Wichita, Kansas, which is one Boeing's largest composites suppliers.

Out of this autoclave come parts like that pictured in the following figure. These are all carbon/epoxy monolithic structures, which means that the entire section of the fuselage is made all at once. This required a considerable investment in automation, like robotic layup and precision fiber placement machines.

Autoclave for making Boeing 787 fuselage sections (reproduced by permission of *Composites World*).

First composite fuselage section for the Boeing 787 Dreamliner (reproduced by permission of *Composites World*).

preform. Instead, a prepreg layup is placed in the mold cavity, the mold is closed up, pressure is applied to consolidate the prepreg stack, and resin is pumped into the heated mold cavity to apply pressure to the prepreg stack, sort of how an autoclave applies pressure to a ply stack. Then the part is cured in the mold before the mold is taken apart and the part released.

Another wrinkle on this theme is what is called the balanced pressure fluid molding, or *Quickstep* process. In this process, the mold is either rigid or semi-rigid mold and floating in a heat transfer liquid to which pressure can be applied. This liquid is pressurized to consolidate the part in the mold, then it is heated to the cure temperature of the resin system. This curing temperature cycle can be very tightly controlled so that the resin doesn't go into a catastrophic temperature rise if the cure reaction is exothermic. The heat transfer liquid can also cool the part slightly to completely control the cure temperature.

There are many more molding processes, both open mold and closed mold, that are used to make composites. Even plastic injection molding can be used to make short fiber-reinforced parts if the plastic matrix can be made the have a low enough viscosity that it will easily flow through the sprues and runners in an injection mold.

Filament Winding of Composites

What if you want to make something long and round like a pipe? Well, a filament winder can make it for you as long as you can make a mandrel that has the internal dimensions you want. Filament winding has a long history, going back to ancient times where Middle Age cannons were made with iron staves wrapped with steel wire. Some of the original French cannons were made in a similar fashion where a lighter weight, thinner wall iron cannon barrel was wrapped in steel wire. Then the cannon was internally pressurized to yield the iron barrel against the steel so that the steel wires remained in tension. This process, called *autofrettage* was used for the higher breech pressures that the French needed to get more range out of their cannons so that the cannon barrels wouldn't split open when they fired them off.

Modern filament winding came about in 1947[9] when the M.W. Kellogg Company was given a government contract to build a lighter weight pressure vessel. Arthur R. Parilla, an engineer, originally developed the method of helically winding a high-strength fiber around a tube with one closed end to make a filament-wound pressure vessel. His 1948 patent, assigned to the Kellogg Company, started the entire business of filament winding pressure vessels. Today, pretty much every solid rocket motor case, and all the liquid propellant tanks for rockets and missiles are filament-wound tubular structures. The liquified natural gas (LNG) tanks that you see on rail cars and trucks are also filament-wound pressure vessels.

In the 1950s fiberglass was just starting to be used to make composites, and polyester resins were coming into their heyday, so most of the original filament-wound structures used glass fiber and polyester resin. All the ballistic missiles built during the early part of the Cold War had fiberglass filament-wound motor cases.

Filament winding machines have two basic parts plus variations on the theme. The part that rotates has what is called a *mandrel* attached to a gear-driven winding head that rotates at a speed set by the operator of the machine. The other part is the fiber lay-down head where the fiber tow is passed through a resin bath to wet it, then kept under tension as the mandrel is rotated. The idea for making a cylinder out of composites using a filament winding machine is that you wind back and forth in a helical pattern at a particular lay-down angle of the fibers. If you control the speed of rotation of the mandrel as well as the speed at which the fiber head moves along the mandrel, you have complete control over the fiber lay-down angle.

Most pressure vessels use either a +-22.5 degree wind angle or a +-30 degree wind angle, depending on whether it is the hoop direction of the cylinder that needs more strength, or the longitudinal direction of the cylinder that needs more strength or stiffness. With modern automatic controls for the winding machines, the wind angles can be changed during the winding process to lay down a few layers at one set of angles and the next few layers at a different set of wind angles. The wind angle can even be varied along the length of the pressure vessel. This gives the tube or pressure vessel designer tremendous flexibility in how their filament-wound cylinder comes out.

Filament-wound structures can also be all composite, or in the case of a pressure vessel, can be lined with a thin aluminum or other metallic liner. Making a high-pressure gas cylinder (like a scuba tank) usually starts with an aluminum liner to make sure that there is no gas leakage through the composite. Then an overwrap of carbon fiber with epoxy resin is wound onto the aluminum liner. Usually, the liner is slightly pressurized during this process so that the liner doesn't bend or deform under the high tension held on the fiber tow. Once the windings are all applied, the vessel is taken directly into an oven and the composite overwrap is cured. If aluminum is used as the liner, it is important to choose an alloy whose aging temperature is quite a bit above the cure temperature of the resin being used. Then once the resin has cured, the pressure vessel is subjected to an autofrettage cycle to yield the aluminum and put the fibers in tension. Then your new high pressure gas cylinder is ready to use.

Filament winding machines come in various sizes based on the length of the winding bed, with some as long as 100–200 feet or more. The 100-foot machines are used mainly for making offshore drill pipe, where two 40-foot-long sections of drill stem can be made at once. Longer, larger diameter winding machines are used for solid rocket motor cases where the rocket is large, like the new SpaceX Falcon 9. This thing is 12 feet in diameter and the first stage is nearly 130 feet long. More typically, especially for smaller composites shops that don't have the floor space, a machine with a 10- to 15-foot bed is common. These smaller machines have a bit more precision than the larger and longer machines, but they basically all work in the same way.

And, again, there are variations on a theme. There are tumble winders for making more sphere-like shapes out of composites, flat winders for making circular structures that are fairly short but very large in diameter, and several variations based on what the end product form needs to be. This is another area where composites designers and clever mechanical engineers have gotten together to come up with whatever they need to build something useful.

Pultrusion

One more process to learn about a little bit before we end this chapter is a technique called *pultrusion*. This is exactly what it sounds like. It is a cross between an extrusion and a fiber pulling machine where the fiber and matrix are pulled through a die to form a long prismatic structure—like an I-beam or a channel. Since pultrusion is very similar in nature to extrusion of plastics and aluminum, you can use it to make any product form that can be extruded. Typically, these products are made using glass fibers and room temperature polyester or vinyl ester matrices, and you can make angles, channels, tubes, basically anything that you can pull through a die. There is also not much limiting the size of the pultruded product other than the size of the pultrusion equipment. Box beams as large as 12" on a side and 1" thick have been made this way.

There are several parts and pieces of equipment involved in this process. First, there is the fiber handling equipment that consists of many *creels* or wound fiber spools all arranged so that they can provide fiber directly to the pultrusion die. This fiber is typically pulled from the center of the creel so that nothing on the fiber end has to rotate. And when a creel is close to running out of fiber, the next creel can be brought up behind it and the end of the fiber from the first creel can be tied to the beginning end of the fiber from the next creel. This way the process can be continuous.

Then there is the resin handling equipment. This is typically a dipping bath where a tub of the resin is kept full, and the fibers are directed through the resin in the tub to completely wet them before they are brought to the pultrusion die. Then comes the fiber preforming, where the wet fibers are collected together to form the basic shape that is needed before they are pulled through the die. Then, as the fibers are pulled through the die, they can be heated slightly to kick off the cure reaction to at least gel the resin so that it maintains the desired shape. Once the pultruded product has been pulled through the die and the resin has cured sufficiently, there is usually a saw that cuts the pultruded product into the lengths at which it is sold.

With the advent of heated dies that can cure the resin almost immediately, come variations on a theme based on the fiber preform that is used. With a heated die you can pultrude a cloth form or continuous fiber mat along with unidirectional fiber tows, and as long as you can wet out the cloth or mat before it goes into the preform dies, once it hits the final pultrusion die it will nearly instantly cure the resin and harden up the pultrusion. Some very intricate pultruded product forms, as well as some very large ones, have been made using this technique. Twelve-inch-tall I-beams, for instance, are easily formed using heated pultrusion dies. Both the web and the flanges of an I-beam, as well as the flat surfaces of a square box beam, can have some fabric or mat internally, which will give the I-beam or the square tubing very good bending strength and stiffness. When you bend a piece of tubing, you put stress in to the web or wall of the tubing. The stresses are in both the direction you bend the tubing and across the tubing. When the tubing is square or rectangular, the stress in the top of the bend is mostly tension stress, whereas the bottom of the bend is mostly in compression. The sides are in a state of shear stress.

These hot dies have to be fairly long because you need some dwell time in the heated die to cure the resin. For smaller cross-section pultrusions, the die is 3 to 5 feet long, but for larger sections, the die is much longer and also much more expensive. The die itself needs to be polished to a very high quality surface finish, and most commonly they are hard chrome plated on the interior surface where the die comes into contact with the curing composite. It is critical that the region in the die where the resin is starting to gel be smooth so that no resin sticks to the die. This makes for a bad surface finish on the pultruded product, so tight controls are used to ensure this doesn't happen. Common chrome-plated die interior surfaces last for about 100,000 feet of pultrusion before they need to be re-chromed, so if the die is used extensively and the pultruded product is in high demand, it makes sense to not only have more than one die, but also to have a die refurbishment process that is based on how much product that die has made.

Finally, there are the pullers that actually pull the product through the die, and the cut off equipment that cuts the pultrusion into lengths that can be shipped. The pullers are either a rubber studded chain drive on both sides of the pultruded part, or a set of reciprocating clamps that cyclically grab the pultrusion on the pull stroke and release on the back stroke. Two sets of these can be used to provide a continuous motion of the pultrusion, but that is not absolutely necessary.

Most cut off saws are what are called *flying* because when they are sawing the pultrusion, they are rigidly mounted to the moving pultrusion table so that they move along with it while cutting the extrusion. Then when they are finished making that cut, they "fly" back down toward the pultrusion die to the next cut location and do it all over again. The length of the cut off pultrusion can easily be changed based on how far the saw moves when it is flying back toward the pultrusion die.

Well, this is enough for fabrication techniques, now we need to move on to the mechanics of composites, what a ply stack really means, why the fiber directions are so important in designing composites, and how to design something wonderful. We will also learn why composites design must take into account how you are going to make your part. Because with composites, again, when you make the part, you make the material, and when you make the material, you make the part. It takes a systems engineering mindset to become a good composites designer.

Let's get on with it and see where this whole idea takes us.

6

Brief Introduction to the Mechanics of Composites, Ply Stacks, Unidirectional vs. Fabric

Now that we know what strings (fibers) are available, and which ones we have to work with, and also what glues (resins) we have to apply to those strings, we have a number of questions that arise because we want to actually make something physical that will serve its intended purpose. We need to add in the other branches of our semantic tree of composites so that we can add the leaves and maybe the fruit to those branches. Design with composites requires knowledge of the trunk of the tree (the periodic table of the elements and its implications for the elements that we use to make the strings and glues that we need), and the three major branches of the tree (the string and its properties, the glue and its properties, and the way we put the string and glue together to make a composite). The minor branches that come off these three main branches of our tree start with the design and mechanics of these materials, how we stack plies together, and when to use one product form or one version of string and glue versus another. In this tree, the branches all interact and intertwine with one another.

Now we need to go back to some definitions that were presented in the first chapter. Hopefully you remember strength, stiffness, stress, and strain because now we need to learn about what engineers use to design structures out of composites that will stand up to the abuse that the structures have to endure.

How do we figure out how to calculate the strength or stiffness of the fiber/resin mix that we want to use? What is the difference between a unidirectional ply and a fabric sheet? How do you calculate the stiffness of a stack of plies, all of which are oriented in different directions? How do you choose between different ply orientations within a single stack of plies to get the properties that you want? All these questions are not only valid, but they are also critical to your ability to actually design and make something useful out of composites. Let's start with the basics and build on that until we have at least a basic understanding of the range of mechanical properties we have at our disposal.

Mechanical Properties of Fibers

We really must start here because, after all, the string in composites is what makes the material have the wonderful properties it has. It is, therefore, the first secondary branch we need to understand. The mechanical properties of interest to us break down into three categories: (1) stiffness (elastic or Young's modulus) in the fiber direction and across the fiber, (2) strength of the fiber (usually tensile strength), and (3) density of the fiber. These are generally the most important mechanical properties to know when dealing with composites. They are certainly what you need to know to start your conceptual design. The table below provides some characteristic mechanical properties of several generic fiber types just to get us started.

Characteristic Mechanical Properties of Generic Fiber Types

Fiber Type	Fiber Direction Tensile Modulus (msi)	Fiber Direction Tensile Strength (ksi)	Elongation at Break (%)	Poisson's Ratio	Density (lb/cu in)
High-strength carbon	33	360	1.1	.26–.28	0.0689
High-modulus carbon	54	150	0.2	.26–.28	0.0689
E-glass	10.4	250	2.4	0.2	0.092
S-glass	12.5	360	2.9	0.2	0.091
Aramid (Kevlar)	12 to 18	330	2.8	0.37	0.05
UHMW Polyethylene (Spectra)	17	370	3.9	0.35	0.035
Alumina	58	44	0.2	0.23	0.14

A little explanation of what these numbers mean is in order to get all those who aren't trained in mechanics of materials up to speed. Some of this is repeated from Chapter 1, but it does bear repeating at this point in the book because these concepts and mechanical properties are so important to the composites designer.

- **tensile strength**—the number of pounds of force it takes to break the fiber
- **tensile stress**—the number of pounds per square inch a tensile force applies to the fiber or fiber bundle
- **tensile or elastic modulus**—commonly known as Young's modulus, it is a measure of how easily you can stretch the material in one direction or the other. It is calculated by dividing the stress applied to a material (in psi or ksi) by the strain that is measured when the stress is applied. We learned earlier that strain is the ratio of how much the material stretches per unit length. Strain is usually measured in inches per inch, which is a dimensionless number, and in fibers this is a very small number because the fibers are very stiff in tension. That is why their Young's modulus is usually reported in millions of pounds per square inch or msi. The same is true for most metals. For example, the elastic modulus of aluminum is usually given as 10 msi, steel's is 29 msi, and high-modulus carbon fiber's can be as high as 50 msi.
- **ksi**—thousands of pounds per square inch—this is how tensile strength is usually provided in English units.
- **msi**—millions of pounds per square inch—usually used for reporting Young's Modulus
- **elongation at break**—this is measured as a percentage elongation—as in inches of stretch per inch of length of fiber—when the fiber is expected to break. This number is the equivalent of strain to failure in an isotropic material.
- **Poisson's ratio**—this is a measure of how much the material being stretched contracts in the other direction from the stretch direction. Think of pulling on a rubber band and watching it get thinner as you make it longer. If you divide the amount it stretches by the amount it thins down, you get Poisson's ratio. Rubbers typically have a Poisson's ratio of around 0.5, which means that they thin much more when they stretch than does a material that has a Poisson's ratio like 0.2. Fibers don't thin out much when you pull on them.
- **density**—just like it sounds, this is a measure of what a cubic inch of this material weighs in pounds. Notice above that Spectra is the least dense of all these fibers. It actually floats on water whereas all other types of fibers are heavier than water. (The density of water is 0.036 lb/cu in.)

Why are all these properties important, you ask. They define the weight, strength, stiffness, etc., of the composite when they are blended with the properties of the resin. Note also that the strength and stiffness properties are only reported in the direction of the fiber. That's because these are the properties that you use to calculate the strength and stiffness of a laminate once the fibers have been incorporated with a resin into a composite ply stack.

Mechanical Properties of Resins

Now we need to know some similar properties for the glues or resins that we are going to use in our composite. These properties are going to be reported a little differently because the resins when they are pure and cured are really just hard plastics and are isotropic, with the same properties in all directions. They are effectively in the same category as the lignin and pitch that hold wood together, so this is the next branch of our semantic tree of knowledge of composites.

When I call these plastics isotropic, I mean that the tensile strength and tensile modulus don't have a direction because they are the same in all directions. The thing that makes composites not isotropic (called anisotropic, or orthotropic, but we'll get to that later) is the fiber itself. There are other wrinkles to this as well, which will arise when we learn about strength of a laminate. The strength of a single ply in the direction of the fiber is dominated by the strength of the fiber—the resin in this case is pretty much just along for the ride, and, of course, to hold the fibers together. But when you turn the ply the other way and pull it the direction that would pull the fibers apart, then it is the bond between the fiber and the resin that you need to worry about because the strength of that bond determines the strength across the fibers. These properties are what we are going to need to put together with the fiber properties to get single ply properties. You will see how we do this in a page or so. First, the table below provides some basic mechanical and physical properties for the major resin types used in the string and glue industry.

Characteristic Mechanical and Physical Properties of Resin Systems

Resin Type	Tensile Modulus (msi)	Tensile Strength (ksi)	Strain to Failure (%)	Poisson's Ratio	Density (lb/cu in)
Polyester	0.145	4.8	1.8 to 3	0.44	0.04
Vinyl ester	0.4	11	2.5 to 9	0.33	0.04

Resin Type	Tensile Modulus (msi)	Tensile Strength (ksi)	Strain to Failure (%)	Poisson's Ratio	Density (lb/cu in)
Epoxy	0.6	12	4	0.3	0.05
Phenolic	0.5	8	0.5 to 2	0.34	0.05
PEEK	0.6	16	40	0.4	0.05

We have a new quantity, strain to failure, that looks suspiciously like the elongation at break quantity in the previous table of fiber properties. That is because this is the same property, only expressed in the way that is usually used for an isotropic material, which all plastic resins are until they are combined with fibers. Since these are isotropic materials, they have the same strain to failure when pulled in each direction. One more thing to understand is that we are only noting tensile properties—the reaction of these materials to being pulled. Typically, metals have the same stiffness and usually the same strength whether they are being pulled or pushed—these are called tension and compression. For quite a few plastics this is not the case; they are somewhat stronger in compression than they are in tension. We are not going to make that distinction in this book, but it is important when completing a composite design to pay attention to any major differences in tension and compression properties. Unfortunately, it is difficult in a composite to tease these out and to use them in design until you have made up and tested samples of the composite layup that you intend to use. Often, composites engineers do not take into account the higher compressive strength of the resins because they want to err on the conservative side. This is considered good design practice: to be somewhat conservative in the values you choose for strength of the materials to make sure that when you make the part it will at least meet expectations.

One other thing to note in this table is that the elastic modulus and tensile strength of these resins are much lower than the equivalent fiber direction properties shown in the table of fiber properties. Also note that the Poisson's ratios of the resins are higher (more like rubber) than most of the fibers—with the exception of Kevlar and Spectra, which are actually thermoplastic fibers, hence they act more like rubbers than carbon, glass, etc. When these resins are incorporated into a ply with fibers, the mechanical properties of the fiber dominate the ply only in the direction that the fibers are running. In both the other directions, it is the resin mechanical properties that dominate. This is a very important concept and one that needs to be understood to grasp the mechanics of a ply stack and how to calculate the expected stiffness, strength, and potential failure strength of the composite layup.

Calculating the Mechanical Properties of a Single Ply

Now we get to the meat of this discussion. How do we calculate the mechanical properties that we need for a single ply of unidirectional fibers using the properties of the fiber and the resin that we have selected? These are the orthotropic properties that I noted earlier. In a three-dimensional, completely anisotropic material there are 27 constants that are needed to describe the mechanical properties of the material. Orthotropic materials on the other hand, are symmetric in two directions—in our case, across the ply and 90 degrees to the plane of the ply (through the ply). So there are only 9 constants that you need to know—way easier to come up with than 27, but more difficult than the 2 properties that you need to describe an isotropic material.

Fortunately, there is a technique called *rule of mixtures* that will at least get you close to some of the properties that you need to design a laminate. What this rule says is that you take the percentage by volume of the fiber and of the resin, multiply the fiber property by its percentage by volume, and add it to the multiple of the resin property and its percentage by volume to get the property of the ply. What is needed is the percentage by volume of both the fiber and the resin. And this is where the art form comes in. What you are striving for is the highest percentage of fiber by volume that you can reasonably achieve when you combine the fiber with the resin and squeeze out all of the excess resin. If you are using a prepreg, the manufacturer of the prepreg will give you the fiber content by volume so you don't have to estimate it.

Let's illustrate this with an example. Let's estimate the properties in the fiber direction for a unidirectional ply of carbon/epoxy. And for the sake of making this realistic, let's assume that you can achieve 60 percent fiber by volume. This is typical of high-performance aerospace composites that are made of unidirectional prepregs or tapes. Using our rule of mixtures, let's estimate the density of the ply.

$$\text{Ply Density} = 0.6 * 0.0689 + 0.4 * 0.05 = .0613 \sim .06 \text{ lb/cu in}$$

Knowing the density of the ply is very important because you nearly always have a weight requirement that needs to be achieved.

Now how about the ply elastic modulus in the direction of the fiber, commonly denoted E_1? We can use our rule of mixtures here too, and it works. But first I want to introduce the concept of the classical mechanics of an orthotropic material to demonstrate some fairly complicated situations. This is not as simple as merely using the rule of mixtures. Since the ply is orthotropic rather than isotropic, instead of two constants to describe the mechanical behavior of the ply—as in Young's modulus I and

Poisson's ratio (nu)—we have a total of 12 of them. First, however, we have to talk about the relationship between E and nu in an isotropic material. Poisson's ratio (nu) is related to the tensile modulus I and the shear modulus (G) through the following expression:

$$G = E / (2 * [1 + nu])$$

This is why it is typical to provide just two of these constants when talking about the mechanical properties of isotropic materials.

For our case, where the material is orthotropic, the 12 constants are E in all three directions, G in all three directions, and nu in six different sets of directions. By this, I mean that if we define the fiber direction as direction 1, and the other two directions as 2 and 3, we will have nu_{12}, nu_{13}, nu_{21}, nu_{31}, nu_{23}, and nu_{32}. In a unidirectional ply, since the two directions that are not the fiber direction (transverse directions) are symmetric, the only three Poisson's ratios that we need are nu_{12}, nu_{21}, and nu_{23}. But that still leaves us with 9 things we must know about the ply to determine the properties of the ply.

There is a very good discussion of this in Volume 1 of the *Delaware Composites Design Encyclopedia*[1] or in pretty much any good text on composites design, so I am not going to go into further detail with this in this book, because I am not trying to teach anyone the mechanics and mechanical behavior of composites. I am only trying to introduce you to the concepts and complexity of the job if you try to do this manually. Fortunately, there are a number of tools that are tailored to the design and analysis of composites that have enormous libraries of properties of nearly every fiber and resin system of interest. We will learn about these in a later chapter in this book, so bear with me here while I simplify this so that you understand what you are asking these computer-based tools to do for you.

Back to what you actually need to know from reading this book. Let's calculate E in the fiber direction using the rule of mixtures approach. Remember that we calculate it using the information that we have about the carbon fiber and the epoxy in the two tables above, like the following:

E_1 (ply) = 0.6 * E_1 (High Strength Carbon Fiber) + 0.4 * E (Epoxy)
= 0.6 * 42.7 * 10_6 psi + 0.4 * 0.6 * 10_6 psi = 25.86 * 10_6 psi ~ 26 msi

Why did I put in the ~ to denote approximately 26 msi? We don't know our mechanical properties with enough certainty to use a number with 4 significant digits at this stage until we make up some samples and test them to get actual mechanical properties. These are the numbers that composites designers use when they are first completing a preliminary design and just need to get close enough that they aren't going to fool themselves. At this stage in any composites design, just getting ballpark numbers is good enough.

Now for the Poisson's ratios—nu_{12} and nu_{21}. In general, these two are not equal, and what is called the major Poisson's ratio—nu_{12} is generally much greater than its transverse friend, nu_{21} (or nu_{31} which is the same in a unidirectional ply). This is because when you pull on a ply in the fiber direction, the contraction in the transverse direction (across the ply) is almost entirely from the contraction of the fiber itself. Whereas when you pull the ply in tension in the direction across the fibers, they are so stiff that the ply does not contract much in the direction of the fiber.

For our discussion, using the rule of mixtures to get close enough to the longitudinal Poisson's ratio is good enough. Again, that number looks like the following:

$$nu_{12} = 0.6 + nu_f {}^* + 0.4 {}^* nu_r = 0.6 {}^* 0.26 + 0.4 {}^* 0.3 = .276 \sim .28$$

where nu_f is the Poisson's ratio of the fiber, and nu_r is the Poisson's ratio of the resin. This is close enough for our purposes here. Again, there are large online databases of this information, most of which have been incorporated into the design, modeling, and analysis tools that are available today. This was not the case 30 years ago—those of us doing these calculations and composite designs did not have the tools that are available today.

Mechanical Properties of a Ply of Fabric

As we learned about in the chapter on strings, fabrics come in many different weave patterns, and can even come in a tri-axial weave where there are yarns at +-45° angles to the fill yarns. Some examples of these are shown in the figure below.

For our understanding at this time, it is safe to start with a plain weave, what is called *balanced* fabric. The properties of one ply of a plain weave fabric impregnated with resin can be estimated by using the properties of two plies laid at 90° to each other. These properties, like stiffness and strength, need to be decremented a bit because of the curvature of the fibers in the fabric as they pass over each other in the fabric weave. This is called the *crimp* of the yarn in the fabric. Many fabrics use a wound or twisted yarn, so most often there is also a twist to the fibers as well as the crimp. Finally, it is nearly impossible to get as high a fiber density by volume in a normal fabric-based composites, so rather than the typical 60–65 percent that you get with unidirectional plies, using a balanced fabric will only let you achieve about a 50–55 percent fiber density

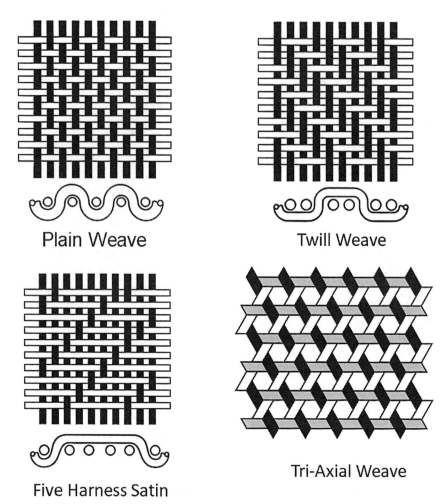

Plain Weave

Twill Weave

Five Harness Satin

Tri-Axial Weave

Samples of different fabric weaves (produced by Candace Hignett).

by volume. There are some caveats to this rule of thumb, but it is at least a good place to start.

As for how much to decrement properties like stiffness and strength due to the twist and crimp of a balanced fabric, a measure that is commonly used is the *efficiency* of the fabric weave that you are using. The more crimps there are per unit length of yarn in the fabric, the lower the fabric efficiency. For instance, plain weave, balanced fabrics have a lower fabric efficiency than do twill or satin weaves. These fabric efficiencies can range between 75 percent and 95 percent. Some satin weaves, like 8-harness satin, can have fabric efficiencies in the warp direction, where the fibers are straight for most of their length, that is nearly as high as a unidirectional ply, or about 98 percent efficient. However, they suffer in the fill direction because there just isn't as much yarn there.

Brief Introduction to the Effective Properties of a Laminate

Now that you know about ply properties, and why they are the way they are, I need to introduce the idea of *effective* properties of a laminate. This is what the composite designer really needs to get started designing whatever they want to make. Essentially, you need to calculate a set of mechanical properties for the entire laminated ply stack based on the fiber, resin, and orientations or fiber directions of each ply. Of course, you have to make a good guess at the fiber volume fraction that you are going to achieve, but those are well known for both unidirectional composites as well as fabric composites. As I stated above, if you are using prepreg, the manufacturer of your prepreg will give you that information.

So far we have discussed what happens at the microscopic level with the fiber and the resin, but to get properties that you can design with, you need to pull back to look at the layup from a macroscopic level—essentially as a monolithic orthotropic material. There are three layup types where calculating these monolithic properties are relatively straightforward. Two of them are easy to understand: a stack of balanced fabric with every other ply turned 45°, and a discontinuous mat layup where the fibers are at random orientations. The third is what is called a quasi-isotropic layup. This is where the fiber stacking sequence goes 0°, +45°, -45°, 90, where each of these numbers is the angular direction of the fibers in the stack. Another way you can make a quasi-isotropic laminate is to make a stack at 0°, +60°, -60°. To make a truly quasi-isotropic layup which is called a balanced layup, you need to make a ply stack that has twice as many plies as the base quasi-isotropic stack, meaning an 8-ply stack or a 6-ply stack. And the second set of plies needs to be a mirror image of the first set. These two-ply layups behave as if they are an isotropic material in the plane of the panel that you are making. This is because if you pull on the 0°, +45°, -45°, 90° layup stack in the direction of the 0° ply, it has a tendency to warp toward that ply because the +45°, -45°, and 90° degree plies are more compliant than the 0° ply in the direction of that ply. The figure below gives a couple of examples of what these ply layups look like. Shown are the unidirectional ply stacking sequences that are quasi-isotropic, and two different fabric layups, one of which is truly quasi-isotropic, and the other one, which does not have any 45° plies, is only isotropic in the two fiber directions. Across the diagonal of this layup the properties are different than

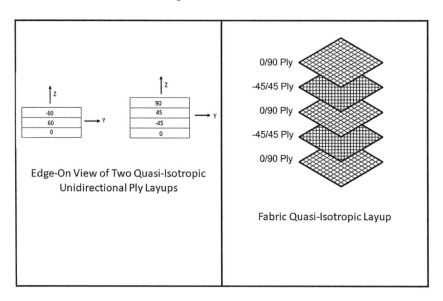

Examples of quasi-isotropic layups, unidirectional and fabric (produced by the author and Candace Hignett).

they are in the two fiber directions. If all your stresses are in two directions in a thin panel of composite, this lack of 45° plies works just fine.

With all three of these types of layups, you can safely assume that at least for thin sections like a car door or airplane wing skin, you can use a set of isotropic material properties for the initial design concept to get you in the ballpark. And if your section is so thin that you can't afford 6 or 8 plies, then it might be best to use a couple of plies of fabric to make the panel, that is if you want to make it quasi-isotropic. If, on the other hand, you want to make something that has specific properties in specific directions, this rule need not apply. Most composite designs end up being thicker than two plies, because the plies are between .005" and .010" thick, and the majority of the panels you want to make are considerably thicker than this.

When we get to other layups for different applications, like for composite overwrapped pressure vessels, or wound driveshafts for race cars, helicopters, some high-end cars like the Nissan 350Z, and even wind turbine blades, the layup gets somewhat more complex. The original method of coming up with these properties is what is called *laminated plate theory*. In today's composites business, there are a number of computer tools available that will perform these calculations for you, but it may be instructive to provide the basics of the theory here, along with resources so that you can easily find this information online. One free resource is the

Composites Design and Simulation Environment from the University of Delaware.

Another free tool for calculating the effective laminate properties is eLaminate from ESP Composites.[2] Their free tool is an Excel spreadsheet into which you can input ply properties and your ply stack layup, and it calculates the elastic properties that you need for design. There is an example already there for you so you can learn how the software works, as well as a place to input your own values. As an example, they start with typical unidirectional properties of a carbon fiber/epoxy ply in a hot, wet condition (~50 degrees F below the softening temp of the resin and 100 percent humidity with water condensing), which is a common place to start because these will give you conservative properties for strength and stiffness. The following figure shows their example material properties, along with allowable strains and strengths. We will go into more depth on allowable stresses and strains in a page or two, but the basic idea is that when you analyze the composite for your design you need to know how much strain it will reasonably take, then decrement that by a certain percentage to ensure that your design works and doesn't fail prematurely. Here is what eLaminate offers as an example of the ply properties needed to get the laminate properties you can use for your design.

From this they have a ply stacking sequence shown in the following figure that is a well-balanced laminate, but it has one extra 0° ply in the middle.

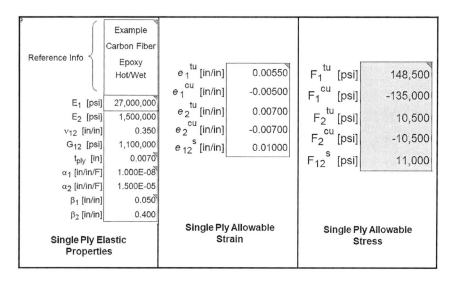

eLaminate sample ply properties (image from ESP Composites eLaminate Software; eLaminate 3.3.xlsx).

Ply (Layer) Number	Mat ID	θ [deg]	z mid [in]
1	1	45	-0.028
2	1	0	-0.021
3	1	90	-0.014
4	1	-45	-0.007
5	1	0	0
6	1	-45	0.007
7	1	90	0.014
8	1	0	0.021
9	1	45	0.028

Ply stacking sequence for eLaminate carbon/epoxy example (image from ESP Composites eLaminate Software; eLaminate_3.3.xlsx).

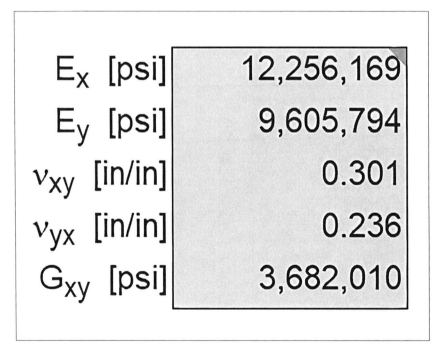

E_x [psi]	12,256,169
E_y [psi]	9,605,794
ν_{xy} [in/in]	0.301
ν_{yx} [in/in]	0.236
G_{xy} [psi]	3,682,010

Effective elastic properties of the subject carbon/epoxy laminate (image from ESP Composites eLaminate Software; eLaminate_3.3.xlsx).

$$e_T \; [\text{in/in}] \quad 0.004500$$

$$e_C \; [\text{in/in}] \quad -0.004000$$

Allowable strains in the subject carbon/epoxy laminate (image from ESP Composites eLaminate Software; eLaminate_3.3.xlsx).

From these two inputs, the spreadsheet will calculate the elastic constants that you need to do your analysis of your layup. They also have a section that allows you to apply loads to your composite to see how it reacts and to determine how strong it is.

Now for the allowable strains, which are really what you are looking for in your design to see how strong the material is and if you are getting anywhere close to breaking the laminate.

Note here that the allowable strains are on the order of 0.4 percent which is typical of how conservative the allowables are in composites design.

About Allowable Stresses and Strains

A bit about allowable stresses and strains is now in order. We use allowables rather than specific tensile strengths like those used in metals because of the variability in cured or fabricated properties in composites. There are variabilities from batch to batch of both fiber and resin, as well as variabilities within a batch. There are also variabilities in the fabrication process that need to be accounted for—remember, when you make the part, you make the material and when you make the material, you make the part.

There are what are called A-basis allowables and B-basis allowables that are specified for each material system on the market. These allowable stresses and strains are obtained by completing a series of tests of the materials under varying environmental conditions. For example, the allowables for a typical carbon/epoxy system are generally given at three temperatures: very cold at less than -100°C, room temperature, and hot (usually wet as well) temperatures above about +200°C. These are broad ranges and each fiber/resin system uses different temperatures.

To derive these allowables a number of test coupons are made up with the material system and tested to failure in each of the environmental scenarios. As many as 100 coupons are required to establish each allowable because the allowables are statistically based. And there are allowables for tension, compression, shear, etc., as well as moisture uptake, hygrothermal properties, crack growth and fracture mechanics specimens, the list goes on. But to get to the point here, an A-basis allowable range of strains, for example, requires that 99 percent of the samples fall within the allowable range with a 95 percent confidence. For B-basis allowables, those numbers are 90 percent of the samples falling within that range with 95 percent confidence. The confidence levels come from classical statistics where 95 percent confidence means that you have 95 percent confidence that the mean value of the parameter you are measuring will fall within the range that you have determined. I realize this is a mind bender for most folks, but since this is all statistically based, we have to use statistics to get us to some number that we have confidence in (sorry about the pun).

Thicker Laminates and the Stiffness Matrix

Something a bit more informative and useful is the thick laminate model, or what one would use to model thicker composites with lots of layers of plies. To understand this, we need to introduce the concept of a stiffness matrix and what that means to the composites designer. I'm not going to get deep into the math of this but I need to introduce the concept and also show a solution for thicker laminates that works. The stiffness matrix is what has to be set up in order to do a finite element analysis (FEA), which is really the only way to reasonably analyze thick section composites and get results that can be used for a design.

What exactly is a stiffness matrix? In basic terms, it represents the relationship between stress and strain for a particular material and geometry. It is represented by the [K] matrix below:

$$\{\sigma \text{ (stress)}\} = [K \text{ (stiffness matrix)}] * \{\varepsilon \text{ (strain)}\}$$

I am using vector and matrix notation here just to make sure you understand what these things mean. In the above equation, $\{\sigma \text{ (stress)}\}$ is the vector of stresses, which in our case are the x, y, and z direction stresses as well as the x-y, x-z, and y-z shear stresses, so this is a 6-component vector. The thing represented by $\{\varepsilon \text{ (strain)}\}$ is the corresponding 6 strains in each of the corresponding directions, three in axial directions, and three in shear. The stiffness matrix [K] therefore is a 6 by 6 matrix of coefficients that describes the relationship between stress and strain. Put together in terms of what we've been talking about, and using the quantities that we have discussed, the stiffness matrix looks like the following for an isotropic material:

$$\begin{Bmatrix} \varepsilon_{xx} \\ \varepsilon_{yy} \\ \varepsilon_{zz} \\ \gamma_{yz} \\ \gamma_{zx} \\ \gamma_{xy} \end{Bmatrix} = \begin{bmatrix} 1/E & -v/E & -v/E & 0 & 0 & 0 \\ -v/E & 1/E & -v/E & 0 & 0 & 0 \\ -v/E & -v/E & 1/E & 0 & 0 & 0 \\ 0 & 0 & 0 & 1/G & 0 & 0 \\ 0 & 0 & 0 & 0 & 1/G & 0 \\ 0 & 0 & 0 & 0 & 0 & 1/G \end{bmatrix} \begin{Bmatrix} \sigma_{xx} \\ \sigma_{yy} \\ \sigma_{zz} \\ \sigma_{yz} \\ \sigma_{zx} \\ \sigma_{xy} \end{Bmatrix}$$

Isotropic material stiffness matrix in terms of Young's modulus and Poisson's ratio (produced by the author).

Why is this important? It is because this is what is used to evaluate the stresses and strains in your composite part when you use an FEA program to apply the loads and constraints that the part will be subjected to. The stiffness matrix for layered composites is not as simple as what is shown in the figure above because composites are not isotropic. Generally, there is a lot of numerical integration that is required to come up with all 36 quantities in an orthotropic stiffness matrix. And, if you had to do this for every element in your model, the computation time would be enormous. However, some simplifications can be made to the generalized stiffness matrix for layered composites. These sorts of things are done by nearly all the modern FEA programs on the market that have solid elements capable of modeling composites. One that I used rather effectively was developed by Dick Christensen and Ed Zywicz of Lawrence Livermore National Laboratory (LLNL)[3], where they used a generalized averaging method to come up two generalized constants, which they called λ and μ where λ is the generalized Young's modulus and μ is the generalized shear modulus. These are based on Lame's constants, but the intent is not to give you a treatise on mathematics at this point, so we will just leave it at that.

Then they established a relationship between these two constants based on the observation that the elastic properties could be separated into matrix-dominated and fiber-dominated properties. Based on the matrix-dominated properties in a single ply, they observed that these properties contributed to the energy stored by straining the material (called the strain energy) as follows:

$$E_{22}/2 : \mu_{12} : \mu_{23} \text{ as } 2 : 2 : 1$$

Where E_{22} is the transverse elastic modulus of the unidirectional ply, μ_{12} is the x-y shear modulus (related to G_{12}) in the fiber direction, and the other (μ_{23}) is his shear modulus in the transverse or matrix-dominated direction. This resulted in the ability to describe the contribution to the stiffness matrix using a single numerical integration for each ply and then summing them up to generate all 36 constants required for the stiffness matrix for the entire laminate.

That's enough of the mechanics of composites for now. Hopefully when we get to design, you will understand the concepts of stiffness and strength of a composite laminate so that you can understand and use the tools that we will cover in a later chapter.

Designing Something Using Composites

Finally, we get to the fun stuff and put some twigs and maybe even a few leaves on our semantic tree of knowledge of composites. How do we design something magical that nobody has ever dreamed up before? Or maybe you just want to learn how to use composites in your next design project and you need a good place to start.

First, some very important things about designing with composites. I realize I'm repeating myself here, but I can't say this enough:

When you make the part, you make the material
AND
When you make the material, you make the part

Composites design is fundamentally about this particular issue. It takes a systems engineering approach: you need to think of how the part is going to be made and how the material is going to be made when you first start planning the design of your part or structure, or possibly the entire assembly that you are going to try to design and build. Most of the decisions about what to use and how to use it are made at the start of the design process. In fact, this is the bulk of the creative part of the design work.

Where to Start—a Good Set of Requirements

While requirements are always the best place to start with any design project, they are absolutely critical for a composites design to be successful. A good set of requirements is essential to coming up with a successful design. What do you want to accomplish? Light weight, high stiffness, direct replacement for a metal structure? Using a good set of requirements that is tailored to what actually has to be accomplished, rather than just using composites because that's what you want to do, will turn your design from a turkey into something that is useful.

To get started you must ask yourself or whoever you are doing the design for several questions:

- What is the application, and what is the design trying to accomplish?
- Who is your customer and why would they want you to use composites?
- What are the restrictions or requirements for:
  ◊ Weight?
  ◊ Strength, stiffness?
  ◊ Dimensional tolerances?
- How is the composite part connected to the rest of a structure?
- Is what you're designing the structure itself? Will other stuff be connected to it?
- What temperature performance does it have to have—highest, lowest?
- Is fire- and flame-retardance needed?
- What are the environmental conditions like humidity, salt, sand, dust, dirt, ice?
- What about transportation, rough handling?
- Is the part or assembly going to be used in a marine or outdoor environment, or indoors in a controlled environment?
- What sort of finish is needed?
- How much should it cost?
- If it replaces something, what is the cost of the thing it replaces, and is there justification for spending more on a composite part than the part it replaces? (Though not always true, it is true in most cases.)

The list goes on. The important thing to answer for yourself is what you are actually trying to accomplish, and why would you select composites to accomplish it. Composites are very useful and wonderful materials and they can make incredible things, but they can be somewhat more difficult to use and are a bit more expensive than metal or other unreinforced plastics, so you have to have a reasonable need to use them.

Once you have a good set of requirements, you know what environment the composite design will be used in, and you and/or your team know why they are using composites, a real design can begin.

What to Do—Makeup of a Composite Design

First, composites design starts just like any other design project, with the exception that it needs to begin from a more fundamental basis than typical metallic design. This is because at the outset of the design, you need to take into account not only the material you will make the composite out of, but also how you are going to make the material. Remember, when you make the material, you make the part and when you make the part, you make the material. The first thing you must do is to decide on at least a rough geometry. As in, how big is it? Is it round, flat, curved, or some arbitrary shape? Usually this is pretty much determined by what the composite part attaches to and what function it performs. This is what can be called the *geometric envelope* for your part.

Once the base geometry has been established, we need to take into account what the part should weigh and what the loadings are on the part. By loadings I mean all the loads that the part has to withstand in its lifetime. The estimated weight and stiffness of your composite will be based on the calculations that you do to get mechanical properties of the material systems that you intend to use. These numbers can be ranges, or things like desired weight versus some threshold value above which it isn't worth it to use composites. Or maybe we have some leeway on stiffness or strength, so we need to develop a desired strength or stiffness range based on the materials that we have available to use and trade against one another. Then we need to take into account where the part or assembly is going to be used because some composites are best used indoors, and some are good to use in very harsh environments. We need to assess these things first because they help us determine what material systems to select as our first good guesses for the thing we are designing.

This is an iterative process because of the nearly infinite number of choices we have for materials, fabrication processes, product forms (unidirectional or woven, etc.). Within this initial design process, we also need to decide on or develop some means of both in-process inspection while we are making the part, as well as inspection of the part before it is put into service, and once it is in service. Trade studies are usually needed at this point before we even have what we can call a good preliminary design. And we also need at this point to leave our options open for different composite material systems, resins, fibers, product forms, fabrication processes, etc., as much as possible. Choosing a fiber or resin or fabrication process at the outset and sticking to it just because you chose it to start with is generally a good way to have a failed design because there are some unknowns in any design project—especially with composites design— that will come back and bite you later if you close off potential options before you have a good preliminary design that will meet all the known requirements. Design margins also need to be inserted into this design scenario as soon as it is practical to do so. An example of this is margin on weight. You may want to use a somewhat less expensive material or a simpler manufacturing process, both of which usually cause the part to weigh more than the more expensive or difficult to fabricate systems. You will need margin on weight in your initial conceptual design in order to trade weight versus cost later in the process and before you get to a preliminary design that will meet the requirements. Another example is margin on strength and/or stiffness. Unidirectional layups are usually stronger and stiffer than using fabric, but fabric is easier and less expensive to use. And some fibers and resins lend themselves better to a high strength design or high stiffness design than do other fiber/resin systems. This is where understanding the mechanical properties and what those properties mean is critical to your design.

Once all this is taken into account in your conceptual design, you have done the iterative process of refining the concept, and done trade studies to get to a geometry, material system, and manufacturing process, the real preliminary design process can begin. Now comes the real work of developing a design that will meet all the requirements. To end up with a good preliminary design, given that you have selected geometry, material system, and manufacturing process, now you need to create a model of the part or assembly that you are planning to build that is true to the mechanical and physical properties of the part and the material. What I'm suggesting here is a 3D design model using a good solid modeling system that will also allow you to work on the part entirely in the model space. As it turns out, composites design is best done on the computer thanks to the burgeoning field of what is called the digital representation of a design, which is basically creating a virtual representation not only of the part, but also the loads on the part, how it is attached to other parts, the fabrication process for the part (tooling, molds, etc.), and the environmental conditions that the part will be exposed to.

The tools for accomplishing this are many and have improved substantially over the last 20 years or so to where it is now entirely feasible to do nearly all your design in the model space. In the late 1970s when CAD was first introduced to mechanical engineers, there were several programs that were available for doing analysis of the design (FEA, dynamics simulation, etc.), but they were not tied in any way to the new 3D CAD systems that were starting to be used in the aerospace business. By the end of the 1980s, nearly all the major FEA vendors had some means of accepting data from any number of 3D solid modeling programs, and sending back results of stress analyses directly to the CAD programs. Since then, there has been a standardization of the languages and the data structures for 3D representations of parts so there is nearly a seamless transport between 3D CAD and analysis programs. But we will get to all this in the next chapter.

Once we have a good preliminary design and have done enough

analysis of that design to be comfortable that it will work, inevitably there are some questions that need to be answered before the design can be completely detailed. This evaluation phase can take many forms, but usually the questions center around: will this work if I make it this way? This is where small test samples and a good test program come into play. What you are trying to figure out is whether some critical aspect of a design, given that you make it with these materials and in this fabrication process, will actually work and survive the environment that the part will be subjected to. It is also important to create samples of the material in the layup that you plan to use and to make them using the desired fabrication process. These small test coupons, when tested to destruction, will not only provide you with actual material properties like strength, stiffness, and weight, but also demonstrate the margins that you have in all those categories.

Now what we have is a good preliminary design that we know will work. And we have selected a material system, a manufacturing process, and have a detailed geometry, including how this thing we are designing either attaches to something else, or is connected in the real world to whatever platform or purpose it is designed for. It is time now to create a detailed design of the part or thing that we want to make including the small details of bolted and/or bonded connections. We also need to do a complete review of the stress analysis and update it with each new detail, and make and test larger coupons of the critical parts of the design that we are pretty sure will work, but need objective evidence to know that they will actually work. This process takes time and is, again, iterative. Doing these larger scale tests will teach us things that we did not already know about the material, the design, or some detail that we are concerned about. Once the results of these tests are obtained, we need to go back and update our detailed design and analysis to make sure that something else didn't get screwed up. In addition to this, we need to develop a quality assurance plan, a reliability plan and model, and plans and documents for all the qualities necessary for the design. And by now we should have also proven our inspection techniques and processes, both in-process while we are making the part and periodic inspections of the part in service.

You also need to address and understand all your supply chain issues. You need to set up reliable sources for all the materials that you need, as well as line up a fabricator that has the capacity to manufacture the volume that you will eventually need—or has the capacity to expand if needed. That is, unless you are going to make it yourself. We will cover this a bit more in a later chapter when we learn about the business of composites and why there are companies that specialize in development and manufacture of composite material parts and structures. All this documentation is usually needed in a complete detailed design, especially with composites. It all needs to be tailored to the material system, as well as the fabrication or manufacturing process that has been selected.

Finally, once we have completed a detailed design, the design and fabrication of the first manufacturable prototype can begin. Unless you are doing a one-off design, this is a very critical step to ensuring long-term success when designing something that will be made out of composites. And this is also more critical with composites than it is with metallic structures because, and I intend to repeat myself here: when you make the material you make the part, and when you make the part, you make the material. See, I told you that I can't repeat that enough when covering composites design and fabrication. And it is, of course, critical that all the composites design, fabrication, etc., work is done within a systems engineering framework.

The prototype that you fabricate should be as close as possibly to what you are eventually going to manufacture. This is to ensure that not only the design and analysis, but also the manufacturing or fabrication process that has been selected, are sound and will work for the entire production run of the parts. This means that all the tooling, jigs, fixtures, resin pumps, molds, etc., need to be production quality and need to be used to build the first prototype. This prototype needs to be extensively tested in a real-world environment and subjected to all the abuse that the part may see in its lifetime. Only then can you go on to either manufacture directly, or to an initial low-rate production if you are designing something that will be made in large volumes.

The figure on the next page demonstrates the entire composites design process, from your initial ideas to manufacture, including all the steps along the way.[1]

What to Do with This? Some Examples

Now that we have all this information and a design philosophy, what do we do with it? The best way of answering this question is to provide you with some examples of design/analysis and technology development projects that I have undertaken over the years. Each of these has been selected to demonstrate the breadth and depth of potential applications of composites, along with the challenges that you face when trying to do something that hasn't been done before. These projects range from development, delivery, and testing of a complete system; to a proposal to develop a technology in the energy industry that would save an enormous sum of money and would benefit the environment in which it is employed;

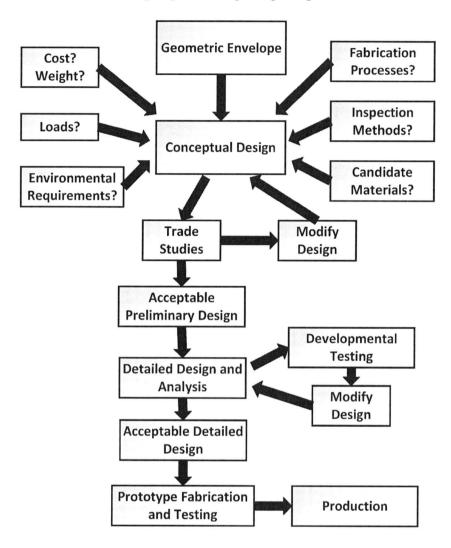

Composites design process (produced by the author).

to a non-destructive testing methodology for aging composite aircraft to give non-destructive evaluation (NDE) technicians a tool to identify what an indication in an ultrasonic inspection means.

Towed Vertically Directive Source— Ten Pounds in a Five-Pound Sack

I'm going to start with a typical Navy project, where they had a need that they couldn't fill with a traditional metallic design. The folks at the

Naval Undersea Warfare Center (NUWC) who are responsible for sonars and submarine hunting wanted an active, high-power sonar array that could be towed by one of their research ships in the Mediterranean—the NATO Research Vessel (NRV) *Alliance*. NUWC was looking for a scientific tool to study the problem of sound reverberation off the sea floor to try to refine their sound propagation models in the ocean and enable them to recognize features that were available from hydrographic surveys. They wanted a low-frequency array (5 transducers @ 600 Hz) and a mid-frequency array (two transducer *staves* @ 3000 Hz) on the same tow body, and both arrays needed to be oriented vertically. Their problem, and why they came to us, was that the winch capacity on the NRV *Alliance* was 8000 pounds, their sonar transducers weighed in at 6000 pounds plus cabling and wiring, and an additional 500 pounds of equipment needed to be on board the tow body. An additional problem was that the arrays needed to be oriented vertically to enable them to paint the bottom with sound, but the deployment ramp on the NRV *Alliance* had been built assuming that it would launch something oriented horizontally. This meant that some part of the large vertical array had to swing up into the tow body. And the tow body had to have controllable flight surfaces so that they could adjust the pitch and roll of the tow body while underway. And, they wanted to tow this thing at up to 16 knots, so the drag forces on the tow point were measured in tons rather than pounds. This thing needed to be strong, stiff, and yet the structure to do this could weigh no more than ~1500 pounds, including the connection to the tow cable. Finally, at least three of the large, low-frequency transducers needed to swing down and lock into a vertical position when launched, and then swing back up into the body of the device as they were recovered. As you may imagine from this description, this was a tall order. Like I said—ten pounds in a five-pound sack.

Obviously, we needed to do something with composites, because a metal structure that would withstand the drag forces and have all the moving part complexity required would have weighed several tons more than the *Alliance* could handle. And the control of the swinging arm that housed the larger transducers had to be passive rather than electrically driven because that would have not only limited the weight available for the structure, but also would have had major implications for the tow cable itself, which was already getting too large and heavy for the ship to handle.

To approach this design, we came up with a number of different concepts that might have worked to try to hone in our ideas into a single geometry that would accommodate what NUWC needed on board the tow body as well as how the tow body would be launched, towed, and retrieved. What we eventually came up with was a modular approach that fit into a standard ISO container, with all its cable, instrumentation, and even the data displays, and could be moved to any ship of opportunity that had the capacity to tow the system and capture data.

For the initial structural geometry of the system, we had to decide how to get a vertical array of 5 large, heavy sonar transducers from a horizontal, stowed position into a vertical data collection position. We thought of swinging an entire arm of all the transducers, swinging just a couple of the transducers, making something that opened and closed like a Swiss army knife, until we finally ended up deciding to have two transducers in the tow body and three on the swing arm.

The structure of the tow body was made of S-glass in an epoxy resin, and it had a Spectra/epoxy outer surface. We had a range of materials that we looked at including E-glass/polyester, S-glass/polyester, S-glass/Epoxy, and carbon/epoxy. At the time, we had just heard about this new ultra-high molecular weight polyethylene fiber called Spectra that might make a good outer coating on the tow body because it would have to be launched and retrieved—in and out of the launch chute—several times during its lifetime and would see very rough service.

The material system choices were made exactly in the manner described in the first part of this chapter, where several things are taken into consideration at the outset. Our major consideration was of course weight, but we also needed to worry about the environment that the tow body would be subjected to as well as the fact that NUWC had a limited budget to design and build this thing. We were looking for something that was as light, strong, and stiff as we could reasonably achieve, while also being very rugged and not prone to fracture or breaking up if the tow body hit something it was not supposed to hit. It also had to be launched and retrieved in at least a state 3 sea (4-foot significant wave height) and survive launch or retrieval in a state 4 sea (6-foot significant wave height), so it was going to bang into the launch chute, and we needed to design for that possibility. We worked up a rough design using each of the materials in our range of choices. We chose S-glass/epoxy for the main part of the structure with four layers of Spectra/epoxy on the outer surfaces of the main part of the body and the swing arm.

Then we had to develop the lightest weight structure we could that would survive all the load cases that we needed to survive. We ended up with an oval exterior shape with an open tunnel in the middle for the swing arm, a sail on top of the body lined up with the large transducer array that held the top transducer, a vertical tail with the two mid-frequency transducer staves, and control surfaces at the rear to control both roll and pitch. And between the oval tunnel and the exterior surface we designed in some frames to tie all the composite structure together. The figure below is a photo of the main body during construction with the side panels left off.

TVDS main structure (reproduced by permission of SwRI).

If you look at this picture carefully you will see a rather large metallic part at the front of the tow body. This was made out of titanium (6Al-4V ELI) because that alloy is nearly impervious to corrosion in seawater, and is lighter than any other high-strength metal available. The tow body is connected to the tow cable at the front through the yoke that you see in the figure, and the swinging strut is attached to the lower end of this titanium piece. This was the lightest design we could come up with that would withstand not only launch and retrieval, but also the drag of the swinging strut when being towed at 16 knots. A picture of the completely assembled TVDS hanging with the strut extended is shown on the next page.

I like to show this picture because the guy standing below the tow body gives you a good reference point for how big this thing actually was. That bump just in front of the rear tail housed the electronics can that has all the control and communication systems needed for the tow body as well as attitude, depth, and temperature sensors—i.e., the 500 pounds of equipment that needed to be on board this thing.

Our weight target was 8000 pounds for the entire system, and the development program for TVDS followed a typical path for a design and development program for the Navy Laboratories—vague requirements and lofty goals. They had some initial ideas of what things weighed, but as we got further into the detailed design of the system, they added

Fully assembled TVDS hanging from a crane (reproduced by permission of SwRI).

equipment and things that they had initially forgotten about. Plus, as we were doing stress analyses and developing a final design, we knew that we were never going to make the 8000-pound weight limit, so we worked with NUWC to get the tow body as light as possible while still having all the capabilities that it needed. The final design came in at 8900 pounds dry (in air) and 2500 pounds wet (in the water). Toward the end of the final design, we also had to add some buoyancy to the upper portion of the tow body toward the rear to ensure that it would tow horizontally—making both sonar arrays vertical—so that NUWC could get the data they needed. At the end of the day, they were happy with the system and used it to gather quite a bit of data in the Mediterranean Sea.

Composite Reinforced Natural Gas Pipeline—Can't Do This with Steel

In 1968, when oil was discovered on the North Slope in Alaska, the first thing that the oil companies had to figure out was how to get the oil down to somewhere in Canada to put it into the oil transmission pipeline system. In the early 1970s, in the midst of the 1973 oil crisis, the first Alaska oil pipeline was designed, and the first pipe was laid in March of 1975. This became what is called the Trans-Alaska Pipeline System or TAPS, which has been in service since 1977, and runs nearly 800 miles between Prudhoe Bay up in the Beaufort Sea and Valdez, which is near Anchorage in the southern end of the state.

Typical oil pipelines operate at fairly low pressures, less than 1000 psi, and the oil is usually warmer than the air around it. Oil that comes up out of the ground has every molecular weight hydrocarbon in it, from tar to methane or natural gas. Normally, the lighter components of the stuff that comes out of the ground are either flared off if the concentration is not very high (that practice will end soon with new environmental regulations), or if the lighter stuff has commercial potential, it is stripped off from the liquid component and put into its own pipeline. Up on the North Slope, there are quite a bit of the lighter components in the crude oil stream, and Alaska environmental laws would not let them flare the gas. Unfortunately, there was no additional gas pipeline to put the lighter components of the raw petroleum into, so the gas had to be liquified at the source in Prudhoe Bay and shipped down on tanker ships during the few months that the ice melt on the Beaufort Sea allowed ships to reach the area where the gas was stored.

In the early 1980s the Trans-Alaska Gas System was formed to try to develop a pipeline design that would carry the lighter hydrocarbons from Prudhoe Bay to either Anchorage, like the Alyeska pipeline, or to Alberta,

Canada, where it could be delivered to existing pipelines that could take it down to the lower 48 states. The problem they had was transmitting that much gas over that much distance would require a very large diameter pipe. However, if you take the pressure up to 3500 psi, you can transmit a dense-phase gas that takes up much less volume, and they could get away with a 36" diameter pipeline. In addition, this gas could be transported in a buried pipeline because it can be transmitted at 10°F and won't melt the permafrost. This would make the pipeline much less expensive than the Alyeska Pipeline, which had to be raised above the permafrost on insulated stanchions since the oil coming out of the well was fairly hot.

The only problem is that a 36" diameter steel pipe designed to carry gas at 3500 psi is very thick and heavy, even if you use very high-strength steel. For example, a pipeline made from 100 ksi yield strength steel (X-100 at the time) would need to have a wall thickness of more than an inch. And since steel pipe comes in 40-foot lengths and needs to be welded together on site, this amounts to a lot of weld material, not to mention the transportation costs just to get the pipe sections there. If, on the other hand, you take a thinner walled steel pipe—say a little less than a half an inch—and wrap it on a filament winding machine with a composite material, you can make a hybrid pipe that has about the same wall thickness, but weighs a little over half of what the steel pipe weighs and uses 75 percent less weld filler material and welding. This is because of the nature of the stresses in an internally pressurized cylinder or tube. What is called the *hoop stress*, which is the stress that would cause the pipe to unzip along its axis, is twice the stress that is in the long direction or along the pipe. This is why cylindrical pressure vessels or pipes that burst always unzip along the axis of the cylinder. Reinforcing that hoop stress with an overwrap of glass/epoxy where the glass fibers are wound only around the cylinder (the easy wrapping direction) allows the glass to take up half of the hoop stress, so the stresses in the steel liner are the same in the hoop and longitudinal directions. In addition, with the composite overwrap you can use a lower strength steel that is not only less expensive, but is also easier to weld and has greater toughness (ductility).

When you pencil everything out, including transportation costs to get the pipe to where it is being laid, and the labor to put it together, it makes economic sense to use a composite overwrap on a steel liner. This was the design scenario that I tackled back at the time that the TAGS line was being proposed. This design scenario followed the same pattern and steps that were outlined at the beginning of this chapter. Available composite systems were the same as before, since this was in the 1980s and some of the newer resins and fibers had not yet been made commercially available. Carbon fiber was ruled out nearly immediately because of the cost of carbon fiber at the time, so we were left with E-glass and S-glass and either an epoxy resin system or a polyester resin system. Since we were going to bury the pipe, and since it was expected to last for at least 30 years without having any problems with corrosion, we needed to be very careful about the selection of both the fiber and the resin system. And, we also had to match the stiffness of the overwrap as closely as we could to the stiffness of the steel pipe so that the overwrap would carry its portion of the stress load. Because this pipeline was going to be 800 miles long, cost became one of the over-riding considerations. Fortunately, there was quite a bit of work that we could draw upon from the composite overwrapped pressure vessel world. At the time, the transition from steel or aluminum air bottles both for recreational divers and for firemen's self-contained breathing apparatus (SCBA) units had been well established, and considerable testing had been done on high pressure composite overwrapped pressure vessels. Relying on that basic set of design and testing information, and to make this pipeline affordable, at the end of the day we settled on an E-glass/epoxy system that was the lowest-cost system that also met the strength and environmental resistance requirements.

There is one wrinkle here, however. To make this work, and to get the E-glass/epoxy overwrap to carry its share of the hoop stress, since the stiffnesses between steel and glass/epoxy are so different, the pipe must go through an autofrettage step. Remember autofrettage? Yes, this is a French word. As it turns out the French invented most of what is used in ballistics today, especially the field of interior ballistics. We covered this a littler earlier, but to remind you, they were trying to strengthen their cannon barrels and keep them from bursting when they were fired without having to resort to wrapping steel cables around the cannon barrel the way their enemies were doing. They discovered that if they internally pressurized their gun barrels above the yield strength of the steel at the inner wall of the cannon barrel without over-pressurizing to burst, they could get the inside of the cannon barrel to yield a little bit so that when they released the pressure, the inside was in compression and the outside was in tension. This made a much lighter cannon barrel, which could be moved more easily and since they could fire them at higher pressure, they would have the range they were looking for to lob cannon balls over the walls of castles they were trying to invade. And if the enemies' cannons were not autofrettaged, they did not have the range of the French cannons, so the French could bombard a castle at will, at least until everyone else figured out their trick and started using that process on their cannons. All cannon barrels today are autofrettaged to make them lighter weight so that they can be aimed more precisely and more quickly.

To get back to natural gas transmission line, this last step, the autofrettage step, is critically important not only to the ability of the composite to carry the load, but also to the life and safety of the pipeline. If the composite is made to carry its part of the hoop load, the steel remains under reduced stress in the hoop direction, and it has very little propensity to burst and unzip along its axis. An all-steel pipeline will burst and unzip along its axis for as long as the pressure stays high enough to propagate the growing crack. But if you wrap it in composite fibers that are oriented only in the hoop direction, you have to break all those fibers to get the pipe to unzip, and the crack gets blunted and turned to run in the hoop direction, where it generally stops and doesn't crack any more. This is called a leak-before-burst (LBB) design and is unique to composite-wrapped pipelines and composite-overwrapped pressure vessels.

But how do you prove that this will work? Well, you have to do a little analysis of the situation, and you have to look for potential failure scenarios. This is also part of the overall design process. We needed to look at all the potential environmental hazards, especially for a buried pipeline that travels 800 miles through what is basically no man's land. Part of the pipe is buried under the permafrost, which is constantly in motion with the seasons and has a tendency to break stuff that is buried in it. What happens if there is significant damage to the composite overwrap? This can happen by accident, if someone decides to dig and hasn't found out where the pipeline is, through intentional acts of vandalism, or even, as mentioned above, if the permafrost cracks or breaks or moves enough to cause a problem. Our composite design has to withstand all these potential scenarios and has to show that it will leak before it bursts. This begat a project with our customer to do an analysis of this situation using the tools available to us at the time. A graphic of the basic model made up of shell elements is shown in the figure on the next page.

What we did in this model was internally pressurize this sliced cylinder and then release the constraints on a few nodes along the line at the top of the composite to simulate a tearing of the composite overwrap. This is a non-linear model that followed the evolution of the damage to see if the steel liner would break or stay competent if a small tear was to open up in the composite overwrap. Using this model, we were going to develop a damage severity prediction by opening up more or less of the nodes along the upper edge of the model to see how large a tear would cause a problem with the steel liner in the pipeline.

This is typical of the kinds of design information that is required for something in a regulated industry like natural gas transmission lines. Safety of the pipeline, life of the pipeline, maintenance requirements, inspection requirements, models of damage and damage effects, models of environmental effects, long-term environmental testing … all these are important aspects of the design process for these sorts of composite

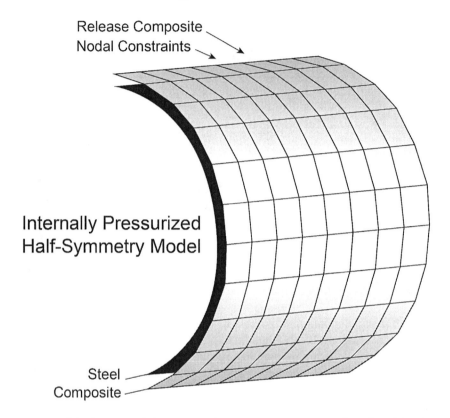

Release Composite

Nodal Constraints

Internally Pressurized
Half-Symmetry Model

Steel

Composite

FEA model of potential damage to the composite overwrap (produced by the author).

designs when the safety of the environment, the local wildlife, and also the people that have to work on the pipeline is involved.

NDE Calibration Blocks—How Do You Tell If It's a Crack or Delamination or Just a Ply Drop? Or Maybe Your NDE Instrument Had a Bad Day?

What happens when you apply a common non-destructive evaluation (NDE) tool to composites? The common tools of the trade are ultrasonic probes and radiographic inspection (X-rays). Radiographic inspections rely on differential absorption of the radiation and have been used extensively on metals and composites. This method can find holes, porosity, inclusions, cracks that go through layers, etc. They are not very good at finding delaminations in composites if the delamination is perpendicular to the X-ray beam. This is where ultrasonic testing (UT) is supposed to shine because the sound waves will bounce off the delamination rather than penetrating through the part. Unfortunately there is a limit to the testing, and this limit is based on the nature of laminated composites, especially aerospace composites, which tend to be made up of many layers of unidirectional fibers in differing orientations.

The most common method of doing an ultrasonic inspection of a part—independent of what the part is made out of—is what is called *pitch-catch*. What happens is that the ultrasonic probe sends a ping of high frequency sound into the part and listens for the echo. In metals, what comes back is whatever will bounce the sound back, which is typically either the other side of the part, or it can be a crack in the material or a void. These are easy to spot in metals because the sound only bounces off discontinuities in the part. With composites this gets a little trickier. Some of the sound can bounce off the discontinuity between layers of the composite. Since composites, especially aerospace composites where the inspections need to take place, have many layers, this makes the signal very noisy and can hide potential flaws. Unfortunately, it is just these types of flaws (disbonds and delaminations) that are critical to making an assessment of the air worthiness of the aircraft. Finally, things get especially tricky in areas like the wing root of the B-2 or the F-22 where a lot of curvature and changing thickness is designed into these areas. Changes in thickness of a composite requires what are called *ply drops.* These are the edges of the plies where the thickness changes, and there are a lot of them in the transition from the wing skin to the wing root. Each ply drop is a discontinuity that will bounce off the sound from an ultrasonic probe. Unfortunately, this is precisely where the UT inspection is focused; it is hunting for delaminations within the wing root itself that could cause the wing root, and therefore the entire wing, to fail under high loads. Definitely something to avoid.

What was needed for this design was some means of calibrating the ultrasonic inspection probes with calibration blocks that have standard discontinuities in them. With metal parts, you can build calibration blocks for finding cracks and discontinuities in metal structures fairly easily. The figure on the next page shows such a set. It is a set of *blind hole* UT calibration blocks that can be purchased online from places like PH Tool. These calibration blocks each have a machined hole placed at a very carefully measured distance from the surface of the block. To use them, a block is stood on end and queried using an ultrasonic probe. The timing of the reflection from the hole is measured and recorded as the distance from the surface to the hole and back for that calibration block. Since the speed of sound is dependent on material, the calibration blocks have to be made out of the same material that is being queried.

Set of blind hole calibration blocks from PH tool (photograph by the author).

With composites, this becomes a little trickier, because not only do you have to match the material, but you also have to match as closely as possible the layup of the part including the ply drops and other discontinuities that are baked into the design. What typically happens is the failed or damaged sections of things like wing roots are cut out of an aircraft during maintenance and/or repair and these sections are used to calibrate the UT probes that find flaws in these critical areas.

The challenge with this one was to develop something like a standard calibration block for composites that could be used in any situation. Again, using the composites design approach, I had to decide what overall geometry I wanted to use, what material, how to make them, what to do to input known flaws, and how to find those known flaws. This one had another wrinkle because of the nature of composites, and because I was trying to come up with a standard way to make these calibration blocks that was as material and design independent as possible. I had to additionally come up with some sort of process of predicting what the ultrasonic probe's response would be to the embedded flaws before actually looking at the calibration block with a probe to confirm that the analytical predictions matched the real world. This required what I like to call a coupled analytical/experimental project where 3D solid computer models needed to be built that simulated the calibration block with known flaws. I used a commercial FEA program to simulate them being probed with an ultrasonic probe and looked at the output data from the simulation to capture the reflection of the sound waves off the embedded flaws. Also required was a design of the calibration blocks with these known flaws that would match the 3D solid models. The intent here was to model the flaw, get the response back from the analysis, build a block just like the 3D model with the flaw embedded, and use an ultrasonic probe on this block to see if the prediction matched the actual ultrasonic data. This would be the preliminary design step of the process. To complete the design of the system, I ran the required the detailed design and analysis cycle: coming up with a design for the block, modeling the block, making a prototype block, measuring the ultrasonic response, and comparing that to the prediction, then modifying the design of the block and going around the same cycle until everything works.

This was a Phase I SBIR project, and I got as far as developing the technique using a 3D solid modeler and a commercial FEA program and building

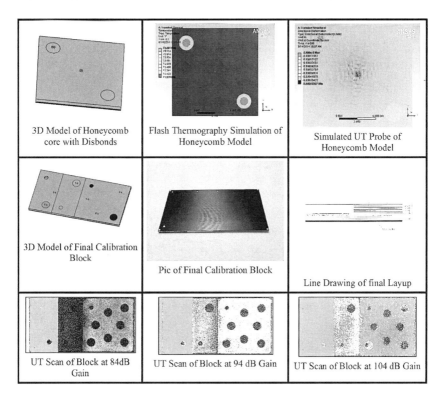

Models, UT scans, simulations, and what was built for a calibration block (produced by author).

a representative block with known flaws in it. The final block that was built had a variety of different layups and included some foam core as well as honeycomb material. In any case, the images on the previous page that were created of the final block, including the analysis results and some 3D model pictures.

Carbon Fiber Bicycle Frames—Go Faster, Win the Tour de France

I want to present one final example of composites design and fabrication, but this time following the evolution of the carbon fiber road bike frame. My intent here it to show how far the industry has come in both understanding this material as well as how to make things that are lighter, stronger, faster, etc. The evolution of the carbon fiber road bike frame is a good example of the increasing complexity of tooling, fabrication techniques, and automation, as well as an example of the complexity of an evolving composite design that started out as a metal replacement and ended up as an entirely new composite design. There is a great YouTube about the history and evolution of the complexity of design, fabrication, and hand layup precision of the modern carbon fiber road bike frame.[2]

Remake of Greg LeMond's 1990 Tour de France–winning bicycle (reproduced by permission of Steve Thompson, The Forza Collection, New Zealand).

The evolution of carbon fiber road bike frames, and carbon fiber bike frames in general, started in 1986 when Greg Lemond won the Tour de France on such a frame. A photo of the 1989 edition of this bicycle (provided by Zach Overholt from BikeRumor.com—with permission) is shown on the previous page.

This bike looks pretty much like every other road bike that you used to see back in the 1980s and 1990s. I show this because it is a good example of some of the original composites designs where the metal tubing in this frame was merely replaced with carbon fiber tubes that had aluminum lugs bonded into the ends of the tubes. The tooling for this was relatively simple because the tubes could be made over a cylindrical mandrel in either a filament winding machine or hand-laid prepreg tape over the mandrel. The forks were a bit more complex, but were made using an internal bladder with carbon prepreg laid over the bladder in a layup designed to provide the correct stiffness and strength in the correct directions. The top tube, down tube, and seat tube were all made using the carbon fiber with aluminum lug method. The rear triangle stays were also made using this technique, as was the front tube for the handlebars and front forks. It is apparent that all the carbon fiber parts were laid up by hand using either pieces of prepreg carbon/epoxy or prepreg carbon/epoxy tape. And all the starting material was unidirectional. This is what allows for the flexibility of putting stiffness and strength only where you need it, making the lightest weight design possible.

Tooling for this bicycle consisted of mandrels for each of the carbon fiber tubes, bladders for the forks, and a heated mold for the forks to consolidate the prepreg. When the forks were placed into the mold, the bladders were pressurized to consolidate the carbon fiber prepreg and push out any retained air to eliminate voids. This sort of process requires quite a bit of experimentation and careful layup of the prepreg to ensure that you don't get any misplaced carbon fibers or air pockets between layers of prepreg, and also to ensure that the temperature and pressure is correct so that the epoxy flows into all the voids but does not completely cure.

Once all the parts come off their respective tools, they must be cured to completely harden the epoxy before the bicycle can be assembled, finished, and painted. Curing typically happens in an oven at a high temperature and in the case of the newer bicycle frames, this temperature is above the melt temperature of the plastic bladders, so they are melted out of the part being cured. The Look 795 Blade RS aero road bike illustrates the other end of the spectrum of complexity in design, tooling, fabrication processes and procedures, and assembly.

This bicycle is, from the ground up, a composite design. The only metal parts in this frame are the crank bushing, front fork bushings, and

Look 795 Aero Blade RS road bike frame (reproduced by permission of Look Bicycles).

rear axle bushings. Everything else is carbon fiber/epoxy. Most of the parts of this bicycle frame are made using the internal bladder with hand-laid carbon fiber prepreg method, where each piece of carbon fiber prepreg is cut using an extremely precise CNC cutting machine. Then each piece of prepreg is hand laid over a bladder that is the correct shape to make the part. The main central triangle is initially laid up in two halves: each half is laid in a mold and then brought together over a folded plastic bladder, then another layer of prepreg is laid over the top of the middle seam to make the triangle. The rear stay assembly is made all in one piece over another set of bladders by well-trained hand layup technicians.

The wonderful thing about this frame is that it uses several different product forms of carbon/epoxy prepreg, depending on the stiffness and strength that is required of the material in each location on the frame. This includes using high-modulus carbon fiber in some places, high-strength carbon fiber in other places, and carbon fiber fabric in those places where the stresses are less well characterized, or where stiffness and strength are needed in all directions. Examples of this are the large structure where the seat tube, down tube, and lower rear stays meet up with the crank. This area takes an enormous amount of bending stress in all directions when the rider is pedaling quickly or climbing a hill. Higher modulus carbon is used in the front fork tubes where stiffness is needed, and you also want to make the fork as light as possible to enhance the maneuverability of the bike.

This bike frame uses more than 400 pieces of precision cut carbon/ epoxy prepreg, each of which is hand laid into a mold or over a bladder in a carefully worked out sequence by expert technicians. The tooling for the two halves of the frame's main triangle is hard tooling because they are going to use this tooling for every one of these bicycles they make, and they intend to make hundreds, if not thousands of them. They also have several heated die consolidation presses that were custom made just for these bicycles. All this equipment is quite expensive, and the fabrication of one frame takes so many labor hours it is no wonder that these frames are $4000 each.

Now you have a more complete picture of not only the evolution of the composites industry, but also the evolution in the complexity of tooling, automation, precision, and clever design that is characteristic of the industry. We have come a long way from direct replacement of metal parts to the all-composite-from-the-ground-up designs of some of the newer, higher-end composite material applications.

As these materials move into high-rate manufacture and price-sensitive applications like the auto industry, more and more of these processes are being automated and greater attention is paid to the cost of the raw materials that make up the composite designs. And, since the auto industry can afford the additional cost of automation because of the thousands to millions of identical parts that they make, they can also afford the cost of high-precision fiber placement or tape laying machines that remove the manual labor from the process. The bicycle market isn't as large or complex as the automotive market, so it will be a while before completely automated carbon fiber bicycle frame manufacture arrives. It won't be that long, though, because the worldwide market in road bicycles has boomed because of the Covid-19 pandemic.

That pretty much concludes this chapter. The remainder of this book will introduce you to the computer-based tools that are available—there are many that are quite good—and the business of composites and where I think that business is going. This last piece is for those of you who are interested in a career in composites, or even enhancing your position in your current job if you work at a company that may have a need for a composite design or to make a composite part now or in the future.

Failure—How and Why Composites Break and How to Avoid It

Now we come to the not-so-fun part. You don't want this to happen to one of your designs.

Or this either (pic on this page).

We do need (pic on next page) to keep our semantic tree together, so I need to elaborate on failure a bit here. Think of this as the intertwining of the other secondary branches and twigs that make the tree stronger.

I need to explain how engineers predict failure in composites and how they avoid it. Since the beginning of the composites revolution that started in the 1950s and 1960s, there have been some spectacular failures along with some spectacular successes, and the industry has learned quite a few lessons from some of the early failures.

Predicting when a composite will fail is much more difficult than predicting when a metal will fail, primarily because composites are relatively

Sinking boat (Shutterstock/Artem Zavarzin).

Broken C141 on tarmac (DepositPhotos/standard license).

new in historical terms, and they are also not isotropic like metals are. With metals, like steel and aluminum, there are well-characterized material properties in handbooks, and the manufacturers of these metals have to meet certain minimum limits on stiffness, strength, fracture toughness, etc., to be able to certify that the metal alloy that they are selling meets these rather stringent requirements.

The same cannot be true with composites because—and I repeat myself here—when you make the material, you make the part, and when you make the part, you make the material. This is true for nearly every composite part that is created or designed. There are a few exceptions, for example you can buy fiberglass tubes, angles, I-beams, etc., but for this discussion we can ignore those. What we really want to do is be able to tell when a composite part that we have designed might fail, and also how it will fail. To bring back an example I offered earlier in the book, when a failure occurs in a metal pressure vessel, it typically rips down its length. This is because the stresses in the hoop direction (around the tank) are twice what they are along the length of the pressure tank. However, when you wrap composite around it, you design the composite to take up the majority of this hoop stress, so if it fails, it usually just opens up in a little area where the fibers surrounding the metal pressure tank break. Remember that this is called a leak-before-burst design, and it is the reason that people use composite-overwrapped pressure vessels. Well that and the fact that they weigh about half (or less) than what their metal counterparts weigh.

The basic problem with trying to predict failure in composites is that failure is always local, and it generally gets initiated at a very small scale—like a few fibers breaking, or maybe a small delamination, or maybe even just a brittle crack in the resin that is holding the composite together, or even a tiny void that was left in the composite when it was made. With metals, there typically are bulk properties that you use to determine where and how a metal part should fail, and since most metals have at least some ductility, you can usually see cracks forming. Sometimes you can even see these with the naked eye, like when a bridge truss develops cracks. With composites, all this initiation and initial growth of the failure happens at a microscopic scale, and you may not be able to detect the flaw until it is too late. This is what we are trying to avoid in our designs, so it behooves us to know before we even start what types of failures there are in composites and what the mechanism is for each type of failure.

But before we start, a couple of notes about how I plan to approach this topic. First, you need to know how bulk composites fail, because the mechanisms of failure locally—at the fiber and matrix level in very small areas—translate fairly well to global failures. Composites failure always happens at an interface of some sort. This interface can be at the microscopic level, as in what happens to both fiber and matrix right at the surface of the fiber or resin, or where they are bonded together. Or the failure can occur at the macroscopic level, as in bonded or bolted joints, places in the laminate where thickness changes and you have ply additions or drops buried inside the laminate, or in tightly curved areas where the bending stresses get rather large. We are going to start with fiber, matrix, and fiber-matrix local failure mechanisms and move on to how those translate to failure at the macroscopic level where you can actually see the failure, sometimes even while it's failing. And then, finally, we are going to learn a little bit about how to predict failure in composites and how to avoid it in your composite designs.

Microscopic Composite Failure Types (Modes) and Mechanisms—How and Why

What does a failure look like at the microscopic level in composites and why does it happen? There have been many books and journal articles written on this subject, and an enormous body of research has been compiled on the aspects of composite material failure. Basically, these microscopic failures, which are nearly always the initiation of the macroscopic failures, can be categorized by where in the material the failure is initiated. These failure types break down along four lines: matrix or resin failure, fiber failure, fiber-matrix bond failure, and what is broadly called *fiber pullout*. This last, fiber pullout, which doesn't happen in continuous fiber composites as much it does in discontinuous fiber composites, is in reality just a combination of two of the other mechanisms. But since it is covered quite a bit in the literature and is something that can be measured rather easily by pulling a piece of the composite to failure in a tensile test, it is usually included in the discussion of composite failure prediction. Discontinuous fiber composites like chopped fiber and glass mat composites can have fiber pullout failures that are neither fiber failures, nor matrix failures. But, since we are only covering continuous fiber composites in this part of the book, we are going to focus on the first three because that is where the initiation of continuous fiber composite failure occurs.

In general, however, failure or fracture of composites at the microscopic level happens when either the fiber or the matrix reach their respective limits in the load that they can accommodate, or the bond between fiber and matrix reaches its limit. While this is also true with metals, in composites it can be the fiber that fails first, or the matrix that fails, or it could be that the mismatch in stiffness between the fiber and the matrix causes too much shear stress (stress along the length of the fiber/matrix bond) for the bond to handle and the fiber breaks loose from the matrix. The latter scenario is the major cause of delaminations in composites where an entire layer of composite comes loose from the layer either below or above it and opens up a crack along the plane of the ply that came loose.

Finally, sometimes it is the environment that the composite lives in that is the root cause of the failure, like when some resins swell up when they are subjected to high heat and high levels of moisture, or even fatigue of the composite if it is repeatedly loaded and unloaded to a level beyond the fatigue limit of the material. These types of failures are usually slow to develop and can be insidious because they can be difficult to detect. And, some composite systems, most notably E-glass/polyester, corrode in the presence of chloride ions and water, as in osmotic blisters on the bottom of a boat that has been in seawater for a while.

Fiber Failure

Fibers typically fail in tension: when they get pulled beyond what they can accommodate they break. Since fibers are typically brittle, this happens without many signs that the load on the fiber has been exceeded. This is because their strain to failure is very small, so when they break they don't release much energy. Unlike metals, brittle materials have little to no ductility, so they don't yield or have a yield strength. When you try to bend a brittle material, it breaks rather than bends. They do have a break strength, and typically they break at a very low percent strain (remember, strain is a measure of how much elongation something has undergone because of the load that is applied to it). Glass fibers have somewhat more ductility than do carbon fibers, especially at higher temperatures and in the presence of high levels of water because the glass starts to soften a bit.

The break site is usually at some point along the fiber where there is a little defect, a discontinuity, or some damage that has happened to the fiber either when it was made or sometime during the handling and fabrication of the part. This is why great care must be taken when laying up fibers before they are infused with resin. Fiber manufacturers try to limit these sorts of fiber defects, and they commonly use what is called a sizing, which is the coating on the fiber that does a couple of things. First, it protects the fiber somewhat from damage, and it also makes the surface of the fiber better adhere to the resin.

Since it is difficult to detect these microscopic flaws, engineers who design composites need to use a statistical approach to predicting failure of their fibers. This is where the A-basis and B-basis allowables come into the picture, and this variability from fiber to fiber, as well as variability along the length of each fiber itself, is the physical basis for why statistics are used to predict failures. It is also why areas of concern for fiber failure are typically places where there are concentrations of stress or strain in a part.

But all is not lost if a single fiber fails because there are so many fibers in each part. Single fiber failure can eventually lead to complete failure, but to do that, all the neighboring fibers that have taken up the load from the single failed fiber have to fail as well. If the neighboring fibers don't have any pre-existing flaws or weak spots, then the single failed fiber doesn't cause much of a problem. The resin can't fail either because it has to transfer the load from the broken fiber to the remaining good fibers. If, however, the fibers next to the failed fiber have flaws, or if the entire laminate in that location is being loaded to near its limit, then when the load from the failed fiber gets taken up by the local fibers, they could fail as well. This is what the composites designer is trying to avoid, and why the allowable stresses and strains on fibers are as conservative as they are.

Resin or Matrix Failure

Resins used as composite matrices are typically less brittle than the fibers that are embedded in them. This is true of phenolics, polyesters, vinyl esters, urethanes, and some if not most epoxies, especially the ones that have been modified slightly to be tougher. When a single fiber fails in a composite, commonly the matrix around it is just fine and is able to easily transfer the load to the neighboring fibers.

However, some more brittle epoxies do develop cracks, sometimes even during the last stages of fabrication of the composite. Every effort is made to avoid this type of failure in high-performance composite parts that are typically high-strength carbon fiber and high-strength epoxy resin, because if this happens during fabrication the part has to be scrapped. The most common reason for this to occur is because the coefficient of thermal expansion of carbon fiber is slightly negative, as in it shrinks when you heat it, whereas epoxies have a positive coefficient of thermal expansion. If the epoxy is brittle, and the composite panel is not designed with this in mind, you might see matrix cracking throughout the part when it is taken out of the autoclave. Good composite designers know this and design around it by either ensuring that they cool the part very slowly, use a more compliant resin, or if that isn't possible, use a post-cure where they hold the part at a temperature lower than the cure temperature for long enough to let the resin do some viscoelastic relaxing so that it doesn't crack. In a sense, this is like firing a ceramic pot wherein you let the kiln cool very slowly or the pot can crack, largely because most ceramics are not only brittle but are, microscopically at least, composite materials.

Another way that matrices crack or fail is fatigue of the part if it is repeatedly loaded to near its load limit. Plastics act in this way much like metals do. If you have ever bent a piece of metal back and forth until it broke, you have seen the effects of what is called low-cycle, high-stress fatigue. If you look closely, before the metal breaks, you can see little cracks opening up on both sides of the place where the metal is being bent. These are fatigue cracks. Resins, which are plastics, do this same thing. If you make a flat, narrow, thin piece of any of the resins used for composite matrices, and bend it back and forth enough times, you will see the little cracks form on both sides just like their metal counterparts.

When you try to bend a composite material that has a brittle matrix like epoxy, usually the first thing that gives a little and forms a crack at the surface of the composite is the resin matrix. It cracks on the tension side of the bend before it cracks on the compression side. This situation needs to be avoided in your composite design. Or, if it can't be avoided, you must use a more flexible matrix material that won't crack when the composite is bent. This is what is done to make fiberglass leaf springs, which are bent repeatedly back and forth as your customized four-wheel drive off-road toy goes over rocks and bounces through dips and valleys.

Environmental factors come into play here as well. Most resins take up some water, especially at higher temperatures. They are what is called *hygroscopic*, meaning that they will absorb a certain percentage of the water that is around them. And when some resins do this, they swell up a little bit to accommodate the water molecules that get embedded in the polymer. If this swelling happens locally and not through the entire composite, it sets up a compressive stress in the swelled area and a tensile stress in the un-swelled area. This only happens occasionally and only with certain susceptible resins. Most other resins get a little rubbery when they are exposed to high heat and humidity, so this is not that common.

Fiber-Matrix Debonding

This type of failure or defect can be difficult to find in a composite until it gets to the macroscopic scale. When the shear stress between the fiber and the matrix gets too high, very small local failures of the bond between the fiber and the matrix can occur. This commonly happens close to transitions and interfaces where the material is changing geometry or the loads are shifting directions a lot. It also happens at edges of the composite where there are free edges and the fiber and matrix at that edge can move somewhat independently.

Fiber-matrix debonding is also associated with both fiber breakage and matrix cracking. When a fiber breaks, the matrix has to transfer the load that was just released from the broken fiber to the neighboring fibers. It also has to accommodate the local change in strain of the fiber that broke. The stress at the fiber-matrix interface now gets very high right at the site of the fiber breakage. This can easily cause the failure of that bond in that local area, and can also lead to further damage because the matrix is no longer able to transfer the load to the neighboring fibers, at least not at that local microscopic area where the fiber broke, so commonly the fiber-matrix bond breaks locally at the microscopic level. This, of course, leads to further damage over time as the composite is loaded and unloaded. The same can happen at the surface where you have a crack in the resin. If the matrix resin relieves tension in the local microscopic area where it just cracked, the stress at the fiber-matrix bond grows much higher and can tend to break that bond in much the same way that the broken fiber caused the local fiber-matrix interface to de-bond.

Another place where fiber-matrix debonding happens is at the edges of things like drilled holes, or sites of impact or other damage that is subsequently loaded up to the point where this interface fails. This is, again, because now you have a free edge, and possibly even some damage locally that causes stresses in the part to require redistribution to other non-damaged areas. Most failures at bolted connections start with microscopic failures of the fiber-matrix interface and then grow to fiber failure, resin failure, and eventually complete failure of the bolted connection.

Macroscopic Composite Failure—It's All About the Interfaces

This is another thing that I can't stress enough. Composite failure that isn't caused by damage almost always happens at an interface of some sort. We saw above how the interface between the fiber and the matrix can fail at the microscopic level. That's your first example. And while there have been many treatises on this subject, all of them follow a familiar pattern. Composites failure usually, if not always, initiates at some interface or another and propagates over time to a macroscopic failure of some sort. So you understand this fully, I need to describe what I mean here and give you examples of how composites fail at particular interfaces.

There are several classes of interfaces, but I want to introduce them in a way that enables someone who wants to design something out of composites (and doesn't want it to break) to not only understand, but also put the information to good use right away. I am going to cover a few different types of interfaces, and how failure both initiates at these interfaces, and how it propagates through the composite.

Bonded Joint Failure

Bonded joints typically fail at the bond itself, at least most of them do. This is partly because the adhesive that is used to bond the two sides of the joint together—whether it is the same plastic as the matrix or not—has to be somewhat thicker in the bond region than in the bulk composite material. Failure is also partly because this interface is being called upon to transfer all the load from one side of the bond to the other through just the plastic or matrix resin adhesive with no help from things like bolts or rivets. Composites can't be welded like most metals can, so they have to be bonded with or without bolts or rivets. This makes them prone to fail. But first we need to see what these joints look like. The figure on the next page shows a graphic of some common adhesively bonded joint designs.

Most of these failures initiate locally and then grow to the point where the joint comes apart. Single lap joints in particular (upper left in the figure above) have several different ways of failing based on how the joint is designed, how well it is constructed, and the strength and ductility

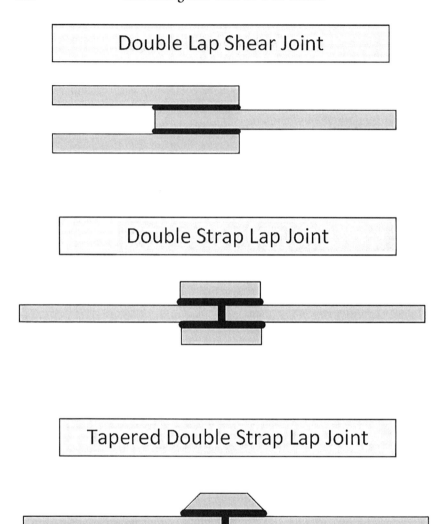

Different ways to make a bonded joint in composites (drawing by the author).

of the adhesive or resin used to stick the joint together. The next figure shows how these joints can fail.

You can see from the above figure that either the adhesive, the interface between the adhesive and one side or the other of the joint, or even the composite layer directly above the adhesive, can fail. Adhesive-to-composite failures like in (a) above are typically caused by poor surface preparation or not using the right adhesive. In the case of the adhesive itself or the

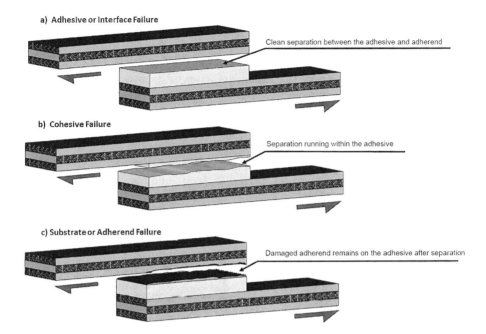

a) Adhesive or Interface Failure

Clean separation between the adhesive and adherend

b) Cohesive Failure

Separation running within the adhesive

c) Substrate or Adherend Failure

Damaged adherend remains on the adhesive after separation

How lap joints can fail (from Omairey, Jayasree, and Kazilas, "Defects and Uncertainties of Adhesively Bonded Composite Joints," *SN Applied Science*, vol. 3, 2021, Figure 4., doi: 10.1007/s42452-021-04753-8, Creative Commons Attribution 4.0 International License).

composite failing, the joint was put together properly, but the stress found the weakest point in the joint and that's where it failed.

When composites and metals are joined together in a lap joint, you can find any one of these three as the cause of the failure. Or as the example below shows in a composite to steel joint, two different mechanisms of failure can happen in the same joint.

Right: **Picture of a composite-to-steel bond where both the adhesive and the adhesive-to-steel interface failed (from Wei et al., "Strength and Failure Mechanism of Composite-Steel Adhesive Bond Single Lap Joints," *Advances in Materials Science and Engineering*, vol. 2018, doi: 10.1155/2018/5810180, Creative Commons Attribution License).**

Bolted Connection Failures

If you don't want your composite joint to fail, add some bolts or rivets to the joint just like you do with metals, right? Probably not the greatest idea unless you do lots of testing and analysis before you try this to make sure that the bolted or riveted joint doesn't fail.

As you can see from these examples of how these things fail, it all starts locally and then progresses from there. The local failure could be caused by almost anything, even a dull drill, like what happened in the photo below.

 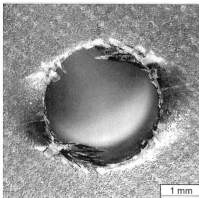

Pictures of a well-drilled hole (left) and a hole drilled by a dull bit (right) that is an initiation site for bolted or riveted connection failure (from Galińska, "Mechanical Joining of Fibre Reinforced Polymer Composites to Metals—A Review. Part I: Bolted Joining," *Polymers*, vol. 12, Figure 6, doi: 10.3390/polym12102252, Creative Commons Attribution License).

It doesn't have to be this way, and you can make very good bolted, riveted, or bonded and bolted/riveted connections if you take the time to design them and analyze them very carefully. When you make the joint you have to pay close attention to how you drill the hole, how well-bonded the structure is, and also how much you torque the bolt. If the connection is made with too high a bolt torque, you can damage the composite at the site of the hole and initiate a failure immediately.[1]

Delamination and Ply Peeling Failures

Yes, these are interface failures too. Only it is the interface between adjacent plies in the composite that fails and not the bulk material itself. Delamination in composites has been studied for several decades, and is

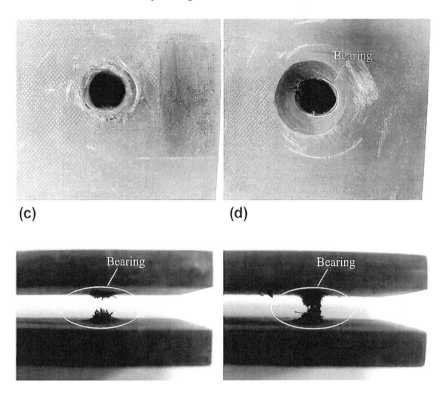

Damage to a composite by overtightening a bolt (from Tobalina-Baldeon et al., "Feasibility Analysis of Bolted Joints with Composite Fibre-Reinforced Thermoplastics," *Polymers* vol. 13, 2021, doi: 10.3390/polym13121904, Creative Commons Attribution License).

probably one of the most ubiquitous ways that composites fail as they age. There is a good description of this on *Wikipedia,*[2] along with a very telling and characteristic picture of a delamination caused by axial (in the direction of the fibers) compression of a carbon/epoxy composite.

There are several causes of delamination in unidirectional continuous fiber composites. It can be caused by bending a flat composite enough that the shear stresses between the layers in the laminate exceed the strength of the resin and an entire section of plies comes loose along a ply-to-ply interface. The result of this will look something like the following graphic.

What happens when you bend a laminate is the outer layers try to stretch and the inner layers have to compress. This sets up a large shear stress between the layers that can overcome the strength of the resin matrix holding them together and the composite delaminates.

Delamination can also happen during manufacture if the part is cooled down too quickly. You have to scrap the part if this happens. Little inclusions between the layers, called foreign object debris (FOD), can also cause a delamination to happen, sometimes even during manufacture.

If the FOD is introduced during manufacture of the part, it will take some loading cycles to get the part to delaminate. This FOD can also be introduced during machining of a composite if the machine tool is dull or the chips from the machining process are not removed quickly enough. The abrasive used in an abrasive water jet machining process caused the FOD to delaminate the carbon/epoxy laminate shown above.

Another potential cause of delamination is damage by an impact on the surface of the compos-

Picture of a delaminated composite compression specimen (from user Kolossos, Wikipedia, Creative Commons CC-BY-SA-2.5 License).

ite, which may also introduce other issues. If the fibers fracture and the composite is still under load, the unloading of the fibers in the layer or ply that is broken can lead to delamination both at the site of the fiber fracture and away from it as the stresses redistribute through the laminate.

Any of these scenarios causes a rather dramatic reduction in stiffness and strength of the laminate, so these scenarios are avoided if at all possible. Fortunately, if the delamination is buried within the composite and it is large enough to cause a problem, it can easily be found with a good ultrasonic examination of the suspected region. The sound waves will bounce off the void where the delamination exists and will give a good reading to the ultrasonic testing device. That is, if the delamination is large enough to show up on the UT scan.

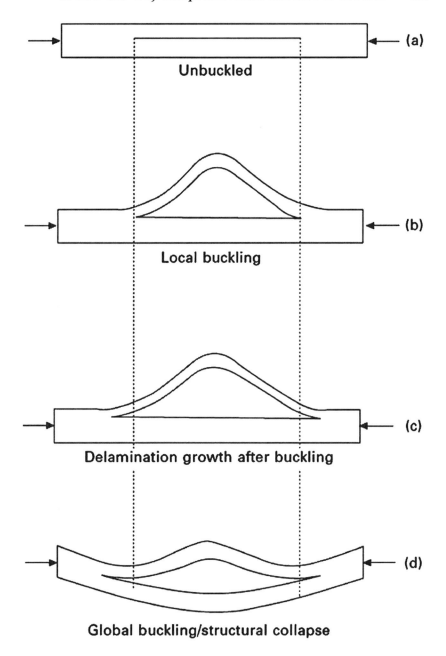

Graphic showing how bending a laminate can cause delamination (From Suriani et al., "Delamination and Manufacturing Defects in Natural Fiber-Reinforced Hybrid Composite: A Review," *Polymers*, vol. 13, 2021, doi: 10.3390/polym13081323, Creative Commons Attribution License).

Failure Initiation at Ply Drops and Section Thickness Changes

Ply drops are quite common in composite designs because of the need to change the thickness of a panel where it has to be bolted or attached to some other part of a structure, or even in areas where there would otherwise be stress concentrations, like where a stiffener or a bracket is supposed to attach.

The small voids at the ends of the plies where they drop off provide sites where there is a discontinuity in stress that can lead to initiation of a failure. This is especially true if the fibers are in the 0° direction, as in the ends of the fibers come straight at the edge of the ply that is dropped off. Since these fibers are not supported at the end, if you pull on the laminate, the only way that the load can get into the ends of the fibers is through the matrix shearing the load into the fiber. Yet again we have a situation where the interface causes a problem—the interface between the fibers in the ply that is dropped and the resin that binds them to the next ply in the stack. If that shear stress gets too high, the interface between this inner 0° ply and its neighbor ply that is still carrying a load will break at the fiber-matrix interface and cause a delamination. That delamination can cause neighboring plies to fail and potentially lead to complete failure of the laminate.

Failure Prediction—How to Avoid Disaster

How do we predict failure in a composite laminate, and better yet, how do we avoid it in our designs? There have been large bodies of work around this topic after some colossal early failures of composite structure, such as pressure vessels bursting well below their intended working pressure and causing significant damage. Even SpaceX has had this problem with a composite-overwrapped pressure vessel (COPV), leading to their rocket explosion in September of 2017. American Airlines Flight 587, an Airbus A300-600R, is another example: the entire vertical tail of the aircraft broke off shortly after take-off on November 12, 2001, and all 260 people on board were killed. The vertical tail on this aircraft is made from carbon/epoxy composite, and it was one of the composite lugs that holds the tail on the aircraft that failed and ultimately caused this disaster.

Predicting these failures during design is the key to making sure disasters like this don't happen. That prediction is typically done by the engineers and material scientists who are designing airplanes, rockets, boats, you name it. They use some classical mechanics and failure theories as well as good design practices, common sense, and a lot of testing of prototypes of their designs.

Failure prediction in composites is done in much the same way that it has been done for metals for the last 100 years or so, with the exception that in composites more focus needs to be placed on the microscopic properties of the material as well as how it is made.

The major failure theories break down along some fairly predictable lines. Unlike with metals where the macroscopic properties of metals like their stiffness, strength, strain to failure, etc., are well known and used extensively in mechanical design, in composites it is the microscopic properties that are more important. The microscopic properties, however, such as fiber properties, matrix properties, fiber-matrix bond strengths, etc., can be related to the macroscopic behaviors and properties that we need to calculate some of the initial failure predictions. These properties are typically related to one another through what are called *constitutive relationships,* where the properties of a laminate are comprised from properties of the constituents that make it up, as well as which direction the fibers lie in a particular ply. We will learn a bit more about this in the next chapter when we cover the computer-based tools that are available to the composites designer and how the finite element method plays heavily into the design and prediction of strength and potential failure of composites.

Essentially, composite failure prediction falls into a couple of recognizable categories: the failure of the bulk composite and local fracture causing failure in the composite part or structure. We will learn more about each category in a couple of paragraphs, but first I need to introduce what I mean by bulk properties failure prediction, fracture mechanics failure prediction in composites, and not only how they relate to the same sort of ideas in design with metals, but also how they are different because of the local, microscopic nature of failure initiation in composites.

Bulk Composite Failure Prediction

There are several composite material failure criteria and failure prediction theories that have been proposed and quite a few have gone through the laborious process of testing hundreds of samples to failure and examining each to see where the failure initiated and how it propagated. The British Government did a rather comprehensive study of this by comparing all the different failure criteria against each other and against real world failures.[3] But they come down to basically three different types, and the choice of which to use is based on not only what you are trying to predict, but also somewhat on personal preference. The lines they break down on for failure initiation are generally stress/strength-based, maximum strain types of criteria, and maximum strain energy (force ∗ elongation) that a particular laminate or composite part can take before some sort of failure initiates. When applying any of these failure criteria, you have to understand that the material properties like break strength, fracture strain, and maximum strain energy are statistical quantities because of the nature of composites. In addition, you have to take into account that when you make a composite you can't make it perfect, so some adjustments of these maximum values have to be applied to whatever criterion you are going to use.

In the past, before we all had Cray supercomputers in our pockets, these calculations were done manually by trained engineers. In today's modern engineering office, computer-based tools including 3D CAD as well as good FEA programs that are tied to them are the norm. The next chapter of the book is devoted to these tools that will help you make these predictions. The best of them have these failure criteria embedded so you can assess your design and predict whether or not some area or part would be prone to fail.

That analysis is by nature an iterative process. It starts with a bulk model of your part in its intended loading condition. The first analysis is usually static and linear. You are looking for are hot spots where you think there might be a problem with your design or the way you are going to make it. As we've seen previously in this chapter, these are areas where there is a transition in thickness, bonded or bolted joints, curvature to the part that is highly loaded, and anywhere that your composite part is connected, or has an interface with a neighboring part. Again, it is all about the interfaces in composites. Looking over your design to start with, with an eye on transition areas and critical interfaces, you will already have a fair idea where there might be a problem. And if you are clever enough, you have modified some of your original design ideas a little bit to take these critical areas into account.

Start with a bulk model of the loads and different load cases that you expect and look over your analysis results to see where there are hot spots. Usually there are a few that you may have overlooked, or areas that you knew about but didn't treat quite as well as you had hoped. In general, at this stage, it is a good idea to modify your design a little bit to reduce the stresses or strains in those areas, if you can.

Next in line is to refine your model and look at your hot spots in detail. You may even have to look at some locations down at the ply level with an FEA program and allow some non-linear displacements to capture what will really be going on in the material if you make it according to your design. Of course, if the loading is dynamic and rapidly changing, you also have to take that into account, especially if the resin that you use is viscoelastic and has a property known as strain rate sensitivity. Lots of viscoelastic materials react much less rubbery when loaded rapidly. Some can even become brittle when you load them very rapidly, whereas if you load them slowly, they can act like rubber.

This is the point in your analysis where you apply your failure criteria of choice. The output from the FEA program will give you all the numbers you need to calculate the propensity to fail, and some of them even have these criteria built into the program. Femap is a good example of this, since it was built to design and analyze composite parts.

Once you have gone through this exercise on a couple of your composite designs, this process will become second nature and just another step in creating a composite design that will work and last.

Fracture Mechanics and Composites

There is a debate that has been going on for at least 20 years about how to apply fracture mechanics to composites, and whether you should use a maximum strain energy release rate sort of fracture mechanics model or one based on the maximum stress normal to the direction of the crack, which is called *maximum principal tensile stress.* There is also the question of whether you should do your fracture mechanics calculations based on the microscopic properties of the composite, which is called micromechanics, or if you should use a macroscopic estimate of the fracture properties of the material and treat it as a homogenous orthotropic material. As it turns out, all these approaches can be used, and to really understand your design and the propensity of the material to fracture, it is often best to use more than one approach to see if your answers are similar.

The macroscopic approach can easily be used to tell whether an existing delamination crack will grow under your load conditions. If you know the fracture toughness of the resin or can do some fracture mechanics testing of a special sample where you have embedded a delamination in the sample, you can determine how probable it is that any delamination type crack will grow. It is probably a good idea to do some dynamic fracture testing along with the static testing, because your resin may be viscoelastic and might get brittle at a high strain rate.

Using a macroscopic criterion like maximum stress or maximum strain for a fracture through the thickness of your laminate is a little trickier. Through the thickness of the laminate, you have to break fibers as well as resin, and these two components of your composite will crack at different rates and for different reasons. As we have seen so far, if a fiber fails under tension, the resin around it has to take up the load and transfer that load into the neighboring fibers. In this case, you may want to use something like a maximum strain energy release rate criterion or approach to predicting crack growth. As each fiber breaks, it releases strain energy as it relaxes and transfers that strain energy to the resin, which then transfers it to surrounding fibers. Using the strain energy release, and the rate at which that strain energy is released as fibers break, can be a very good way to assess how rapidly your crack will grow or if it will grow at all.

There are several books written on this subject that are excellent references, so it is my intent here only to introduce these ideas to you and to give you enough of the vernacular, as well as the physics and material science that you need so that you can understand these more in-depth treatises on this subject. I am also not going to go into the math here because most of that has been taken care of by the companies that provide the modeling and analysis tools of the trade.

Now we go on to the next chapter where I introduce these modeling and analysis tools, where they came from, a tiny bit of the physics, mechanics, and math that went into them, and what all their capabilities are. Stay with me here; you won't regret it.

9

Computer Based Tools for Composites—3D Models and FEA

I've referred throughout the book to how the composites modeling and analysis tools have come of age since the 1980s and are now perfectly adequate—and some may say great—at doing all the math that used to take folks with PhDs in engineering many hours to accomplish. Back in those days the researchers in the DOD and other national labs were doing all the math and building their own home-grown tools so they could use the computers that were available at the time to design and analyze what they wanted to build. I've already mentioned the work of Christensen and Zywicz of Lawrence Livermore Labs and how their math allowed for a more efficient calculation of the mechanical properties needed for the analysis of thick-layered composites. And I should know, because I once needed to build my own code to do this analysis because none of the FEA programs had the facility to deal with thick composites. It used to take a Cray supercomputer several hours to crunch through some of the larger models—that is if you could get time on the Cray.

Let's put some leaves on our tree and get down to business. These are the tools that you will see used today to do composites design. In today's world, these tools can give you answers in seconds, partly because we now have the power of a Cray in the palm of our hands (thanks to Moore's law), and partly because the engineers who work in the large analysis software companies have honed their craft and have graduated from universities that taught them everything they needed to learn about composites. The University of Delaware's Mechanical Engineering Department focused early on composites. There is also MIT, Stanford, Berkeley, Northwestern, Cal Tech, UCLA, Georgia Tech, Penn State, U of Illinois Urbana-Champaign, U of Washington (Boeing, right?), ... the list goes on in today's world. Most of the folks who write the software at the big analysis and design software houses come from one of these universities and have been trained by some of the best minds in the business.

Most of the tools mentioned in this chapter are part of larger suites of analysis software, or they work with one or more 3D solid modeling or CAD tools. And many of them that are specific to composites are linked or can be linked to the major FEA products on the market today. And, since they are commercial software, they are in the business to make money. I'll go into a little more detail about that and the business consolidation in the 3D modeling and analysis world when I get to the business of composites and where I think it's going, but for now I just need to let you know that these commercial products are all very good. Choosing one comes down to personal preference, or what you have available where you work.

Most of the tools available to the composites engineer—budding or not—have great material libraries and are adding new fibers and resins to their databases as the industry creates new products. They also have the ability to add a material or a new combination of fibers and/or resins to let you define exactly what you want and analyze it just as if it were a commercially available material. Most, if not all, of the tools use techniques similar to those developed by Christensen and Zywicz, or at least something that resembles the approach they took. I know from personal experience that John Swanson of Swanson Analysis (ANSYS) was quite interested in Christensen's work because he knew at the time (circa 1990) that he was facing a competitive nightmare if he did not get ANSYS up to speed in composites analysis.

In what follows I am going to attempt to mention as many of the tools as I can that are available to the composites designer. These tools break down into two different approaches to supplying software: they either are stand-alone tools that link to 3D CAD and FEA programs, or they are embedded within or are add-on tools to a particular modeling suite. Again, this is partially because of the consolidation in the 3D engineering software world. The tools that are stand-alone also have trial versions, or in some cases free versions, that you can download and use right away. The free versions typically have somewhat limited capability to do a real design, but they will give you a start and allow the newbie to composites to experiment a bit.

Stand Alone Tools for Composites Design and Analysis

There are a few of these, and more are being developed over time, but a good internet search will bring up some decent tools that are either free for a limited version, or that give you a short trial to allow you to get comfortable with the program before you buy. Added to the mix here are software packages that are purpose-built to just do the hard part of setting up the models and constitutive relationships required to do composites analysis. These latter stand-alone tools are usually available as add-ins to major 3D engineering modeling and analysis programs. There are also purpose-built stand-alone programs that do it all. These programs are usually written and managed by larger engineering software houses whose business model is to create purpose-built tools for particular industries or fields of engineering.

ESP Composites

This is a small one-man company formed by Brian Esp who has a PhD in mechanical engineering and has worked at some of the major aerospace companies. He has also been an adjunct professor at the University of Texas at Arlington, where he taught graduate level courses in composites. He has written a book *Practical Analysis of Aircraft Composites,* about analysis of composites in aerospace, which is, from what reviews I have seen, exactly what it says it is—practical. He has written a set of composites tools that are free to download for the individual user as long as the individual does not intend to make a profit through the use of his tools. His laminate analysis tool, eLaminate, was mentioned earlier in this book as a good example of how to approach designing a laminate using the properties of the fibers and resins as well as the stacking sequence of the laminate. All his tools are Excel-based and most are free, at least for the introductory versions. More advanced applications are available to people in for-profit companies. ESP also has a limited set of finite element analysis programs, with hooks to MSC Nastran, for doing some basic modeling of beams, columns, etc. And Brian is working on developing more tools to fill out his suite of structural analysis software.

Link: https://www.espcomposites.com/

AnalySwift

This is another small company headquartered in Indiana that provides what they call *multi-scale modeling of composites.* Their stand-alone product, SwiftComp, can either perform the structural calculations for you or also provide inputs to your analysis program of choice. This is common with the stand-alone software suites that are available to model a composite structure. They can model a number of different types of composites using their technique. What they do is to separate the model into two parts: a constitutive relationship part and a fairly standard analysis part. This is the way it used to be done by hand, but AnalySwift has coupled these together to allow the composites designer to do all their analysis in one place. They do, however, require you to provide inputs to the constitutive modeling part of the software, and don't have a library of readily available composite materials to choose from. They have relationships with quite a few universities, not-for-profit institutions, and even major commercial engineering software companies. If you are in a university that partners with them, and want to use the software for your own purposes, they will provide it to you free of charge.

Link: https://analyswift.com/

CDS: Composite Design Software—
University of Delaware

The University of Delaware's Center for Composite Materials has their own composites structural design and analysis software tool that has been developed through many years of research in composites and composites manufacturing. University of Delaware has one of the longest standing industrial consortiums in composites in the nation, having survived since its founding in 1978. Consortium members get CDS free of charge, which has an extensive library of materials and will connect to LS-Dyna for non-linear analysis and failure modeling of composites.

Link: https://www.ccm.udel.edu/software/cds-software/

Digimat—e-Xstream

Digimat is a purpose-built materials modeling and analysis program originally authored by e-Xstream in Belgium. It offers a complete package of composites modeling and analysis. They have an extensive material library and have been partnered with material suppliers and OEMs ever since their inception in 2003. They are a major player in composites and materials engineering in Europe, and their software is used extensively by major aerospace manufacturing companies. They are also partnered with nearly all the major engineering and manufacturing software companies. They were acquired in 2012 by MSC Software (NASTRAN), which was then acquired by Hexagon, a multi-national information technology and engineering applications provider. Digimat has remained a stand-alone product just for the design and analysis of composites.

Link: https://www.e-xstream.com/products/digimat/about-digimat

Fibersim—Siemens

Fibersim is a package that specializes in the development of the composites design itself, and has quite a bit of automation of layups and their geometry, materials and material blends, etc. It was developed by a company called VISTAGY, Inc. as an adjunct tool to 3D solid modeling and CAD programs, and it still serves that purpose today. VISTAGY was purchased by Siemens PLM Software in 2011, so now Fibersim is Siemens' composite material modeling program. It is an add-in to some of the most popular and heavily used 3D CAD tools on the market, namely NX (Siemens), Catia, Solidworks, and Pro/Engineer. And since it is now owned and distributed by Siemens, it is the add-in of choice for composites design in NX, which is based on a venerable old 3D CAD system named Unigraphics.

Link: https://www.plm.automation.siemens.com/global/en/products/nx/fibersim.html

Helius: MCT

This software tool came out of the Mechanical Engineering Department of the University of Wyoming, which has been doing original research in structural mechanics of composites for some time. In the late 1980s researchers there were working on failure mechanics of composites when they realized that there wasn't much in the way of software available to do the structural mechanics and failure analysis of composites, so in 2001, they formed Firehole Composites and launched their product, Helius: MCT, in 2009. Most of the original work on the software was done through the SBIR and STTR programs, small business technology development programs sponsored by the U.S. government. The MCT in their software's name refers to what they call *multicontinuum technology,* in which they separate fiber-dominated properties from matrix-dominated properties to develop constitutive models that can be used in FEA. In a sense then, this is similar in nature to Christensen's work. Firehole Composites was acquired in 2013 by Autodesk and now is an application in the Autodesk suite of engineering modeling and simulation software.

Link: https://www.autodesk.com/products/helius-composite/overview

CompoSIDE

This is a somewhat different take on the traditional composites design and analysis software business. CompoSIDE, developed in the UK, is an entirely cloud-based and integrated set of design, analysis, and data management tools, which are independent modules but share the same database. The heart of the suite is an extensive materials library that is accessed by all the other design and analysis modules. There are modules for laminate design, beam design, FEA of beams and shells, a bill of materials generator for your project, and even a yacht design package for sailing yachts and large powerboats. They also have a desktop client for accessing all the cloud-based modules that acts as a repository for your composites design project.

Link: http://www.composide.com/

Composites and Major Engineering Simulation Programs

Most, if not all, of the major engineering simulation and analysis software companies—and I'm referring mostly to 3D solid modelers and FEA programs here—have either acquired composites modeling capabilities or have developed some of their own capabilities. This is true of the majority of the major FEA vendors, who all have a set of elements like shells, beams, and 3D solids that have orthotropic constitutive models built into them. And most of them, if they do not have pre-processing capabilities for composites, have relationships with some of the packages mentioned in the previous section. Here's a brief list of these major engineering simulation packages, which started as FEA packages, and their respective capabilities and add-ons.

Abaqus—Dassault Systèmes Simulia

Abaqus, first released in 1978, was the first product of Hibbitt, Karlsson, and Sorensen, Inc. According to the wiki about Abaqus, Dave Hibbitt and Bengt Karlsson met at Brown University where Pedro Marcal (MARC Analysis Research Corporation) was a professor. They met up with Paul Sorensen when they were finishing their PhDs at Brown in the mid–70s. Working under Pedro Marcal at Brown gave Hibbitt and Karlsson the training and the idea to develop a commercial finite element code that would handle the highly non-linear and dynamic behavior of materials and structures. The first use of Abaqus was for nuclear fuel pin aging simulation at the Hanford Nuclear Reservation operated by the DOE. Originally it was what is called an implicit solver, which is what most linear and non-linear FEA packages are. Later on, two additional packages were released, Abaqus/CAE as a pre- and post-processing tool, and Abaqus Explicit for highly dynamic analysis. Over the years, Abaqus/CAE has added the capability of modeling composites, and both the implicit and explicit version of the program have elements with orthotropic material models that can model composites. One of Abaqus's features is its broad array of materials and innate capability of modeling nearly any material that you can dream up. By the time Abaqus was purchased by Dassault in 2005, Abaqus was already very fluent in its composites modeling capabilities. Dassault has renamed the package Simulia and has integrated it into its suite of other tools, namely Catia and Solidworks.

Link: https://www.3ds.com/products-services/simulia/

ANSYS Composite PrepPost

As I mentioned earlier, John Swanson, founder of Swanson Analysis and primary author of the ANSYS FEA program, was very interested in the composites constitutive model developed by Christenson and Zywicz at LLNL. At the time, engineering workstations were just being developed and were the size of a washing machine. His initial focus was his Prep7 pre-processor, and the post-processor that he had developed to graphically display the results of an ANSYS simulation. Since that time, Swanson Analysis has added capabilities to its pre- and post-processor, as well as new functionality to its suite of elements in the ANSYS program. The basis of his composites constitutive relationships is very similar to the ideas of Christensen and Zywicz. Today, ANSYS Composite PrepPost has very robust capabilities for modeling and simulating nearly any composite structure that can be imagined. And it is completely integrated into the Ansys Workbench simulation integration program. The Workbench suite of engineering simulation tools includes the original ANSYS Mechanical, as well as ANSYS LS-Dyna (Livermore Software Associates), which is a nonlinear dynamic FEA code developed by John Hallquist of LLNL, and ANSYS Fluent (Creare), a computational fluid dynamics code developed initially at the University of Sheffield by a group of professors and graduate students.

Link: https://www.ansys.com/products/ansys-workbench

MSC Nastran/Patran

I noted this above when I covered the Digimat from eXstream. A little history lesson is in order here. NASTRAN was originally developed by NASA in the 1960s.[1] The name NASTRAN stands for NASA structure analysis, and it was originally released in 1968. Shortly after its initial public domain release by NASA, the MacNeal-Schwendler Corporation released its own commercial version of the code under the name MSC/NASTRAN using the original Fortran source code developed for NASA. Since that time NASTRAN has gained in popularity, and there are a couple of other companies that took the original source code written for NASA by the Computer Sciences Corp (CSC). In 1994, MSC bought PDA

Engineering and their software package Patran, which at the time was one of the leading FEA pre- and post-processing programs. Since then, and after several other acquisitions, MSC was purchased by Hexagon. Now both these programs, as well as Digimat and several other software applications, are all part of the Hexagon suite of engineering software tools. The current fully integrated suite is called MSCOne.

Links: https://www.mscsoftware.com/product/msc-nastran and https://www.mscsoftware.com/product/patran

Femap/NX Nastran

Femap was created by George Rudy and his company Engineering Software Products (ESP) in 1985 as a pre- and post-processor for Nastran. It was a competitor for SDRC-Ideas at the time, so in 1999 Femap was acquired by SDRC. And in 2001 SDRC was acquired by EDS (Electronic Data Systems).[2] As you can see, there has been quite a bit of consolidation in this industry over the years and there are only a few remaining large players in the business. EDS spun off all its PLM software, including Femap, to UGS, the original purveyor of the Unigraphics 3D CAD program that eventually became Siemens NX. Femap is now part of the Siemens suite of software tools, and it is tightly integrated with NX Nastran and Fibersim.

Link: https://www.plm.automation.siemens.com/store/en-us/femap/-femap-nx-nastran.html

Solidworks—Dassault Systèmes

Finally, we can't leave this chapter or discussion without mentioning Solidworks and its composites modeling capabilities. Solidworks, founded in 1993 by MIT grad John Hirschtick,[3] has rapidly become one of the most popular and widely used 3D solid modeling programs on the market. The first version of Solidworks was released in 1995 and became very popular because of its intuitive user interface and the ease with which engineers could model what they were designing. It was popular enough that it caught the attention of Dassault Systèmes and in 1997 Solidworks was acquired by Dassault and became part of Dassault's suite of engineering simulation software. It has since sold over 3.5 million licenses worldwide and has established itself as one of the preeminent parametric solid modelers on the market today. Composites design and analysis is embedded in the Solidworks Simulation tool and is completely integrated into the Dassault Solidworks suite. Solidworks Simulation includes modeling of layered composites, has an FEA solver, and can output results graphically using the Solidworks Parasolid geometry engine. It is somewhat limited in scope for thick composites, but for the average composite design, it works well.

Link: https://blogs.solidworks.com/tech/2018/07/solidworks-simulation-an-intro-to-composite-analysis.html

That's about all I'm going to introduce in this chapter. There are other tools and even home-grown and small tools available to the composites designer. There are also more purpose-built simulation and design programs that are available that could be useful if your design fits one of them. But my goal here was to introduce the major players, as well as some up-and-comers that have some good software, so that you have somewhere to start looking for your composites design, analysis, and simulation tool of choice.

Now on to a couple more things. First, I need to introduce you to some other types of composites where the string isn't continuous, and/or the glue isn't plastic. Then, we go on to the business of composites and the world of consolidation in this industry. It's an interesting story and replete with back room deals and intellectual property battles, especially in the fiber and resin supplier business. Then I'm going to provide information about how to make a career out of composites and where you might be able to find a job. Hope you stick with me.

10

Other Types of Composites

What about when the strings aren't continuous? What if the glue isn't plastic? Well, there are other composite material types, and this is where we need to address them a little bit. These are the little minor branches that fill in the canopy of our semantic tree and make it full and ready to bear fruit for us. In general, the other types of composites can be broken down into two distinct groups: discontinuous fiber composites where the matrix can be a plastic or metal, and a grouping of continuous fiber composites where the binder or matrix is not a plastic. This last includes continuous fiber ceramic and metal matrix composites and a class of materials called carbon-carbon composites. I'm not going to delve too much into these last non-plastic matrix composites because they are not used much anymore, except in very special applications, primarily because they are very expensive and very difficult to make. What I am going to cover more are discontinuous fiber or particle-reinforced composites. These can have a plastic or metal matrix binder that is reinforced by hard, stiff, short fibers or hard, stiff, strong particles or whiskers.

Discontinuously Reinforced Composites—
Chopped Fibers and Particles

Discontinuous fiber composites are defined just like their continuous fiber cousins by the fiber that is in the composite and the resin. The main difference is that in discontinuous composites, the length of the short pieces of fiber matters quite a bit in figuring out how strong and stiff the resulting composite will be and how easy it is to process the material and form it into shapes. Shorter fibers produce a weaker and less stiff material that is much easier to process than the longer fiber materials. Typically, these composites in thin sections like sheets or plates are isotropic in the plane of the sheet or plate. That is, as long as they are truly randomly oriented and also short enough in comparison to the width of the sheet or plate. As the fibers get longer and the process for manufacture gets more directional, like extrusion, these fibers tend to align in the direction of the extrusion. If the section is thick enough and the fibers are short enough compared to the thickness, these materials can behave just like an isotropic plastic or metal, albeit stronger and stiffer than their unreinforced brethren.

Plastic Matrix Discontinuous Composites

The most common of these are what are called molding compounds. These come in two forms—bulk and sheet—and are basically a mix of short (1″ or less) glass or carbon fibers in a thermosetting matrix like polyester, vinyl ester, or epoxy. As mentioned above, they are either provided in sheet or bulk form or are mixed on site, typically in an automated mixer where they are mixed together right before they are used. The most common uses for them are compression molding, injection molding, or other ways of making the part that can be easily automated. I think you see where I'm going here. The automotive manufacturing industry uses both bulk molding compound and sheet molding compound extensively, primarily to cut the weight of the car or truck to increase gas mileage. The inside fenders of your newer Toyota or Ford are all made from sheet molding compound, as is the trunk liner, interior door panels, dash boards, you name it. You are driving a car today that is probably nearly half plastic by volume, if not by weight. A lot of that plastic is a discontinuous composite.

Sheet molding compound (SMC) is made in large rolls by a machine that puts down a layer of resin that has hardener already added on a thin plastic sheet (this peels off later) then lays down a layer of chopped fiber, then another plastic peel sheet, and this assembly is run between two rollers that are spaced just the right distance apart to make the sheet the desired thickness. There are challenges in doing this fast enough to make sheet molding compound cheap enough and in enough volume to meet the demand of all the industries that use it. Most of the manufacturers of SMC that have survived have worked out the kinks and details in this process, and can change fiber lengths or resins at will because they have automated all the parameters needed to keep the resins at the right temperature and viscosity, to get an even distribution of the chopped fibers, and to get the right thickness for the sheet without having the resin be so runny that it doesn't make an even sheet or so stiff that the rollers break it as it is being rolled.

This sheet molding compound is then either chopped to size and put directly into the mold, or it is rolled up, leaving one of the two plastic sheets on it as a peel ply so that the roll doesn't stick together in one mass.

If it needs to be stored for some time, it is put in a freezer and held at a low enough temperature that the resin can't harden any further. Then it is delivered to wherever it is going to be made into a product.

Thermosetting bulk molding compound (BMC), sometimes called dough molding compound, is either mixed on site with a continuous mixer feeding an injection molding or compression molding machine that can make hundreds to thousands of molded parts, or it is made into fat cylinders (~4") and immediately cooled down to stop the resin from hardening. Then chunks of this material are chopped to the desired length and loaded into the automated molding machines. Either way, this material is one of the most common materials to use in the appliance and auto parts manufacturing business. BMC is used in lots of housings for electronics, in motor housings for washing machines and dishwashers, in auto parts, and in all sorts of parts that need to be stronger than plastic, water-tight, non-conductive, and resistant to corrosion. You probably have a toaster on your kitchen counter that has a base and frame made out of BMC. Certainly you have a blender, or a food processor, or some other appliance that has a BMC base and frame. You probably didn't know it was made out of a composite material. Your washing machine has both motor and pump housings that are made using BMC.

Thermoplastics are also used to make discontinuous fiber composites. These span the range of fibers or reinforcements from mineral fill to chopped glass fiber, to even high-performance chopped carbon fiber. The thermoplastics also run the gamut, from polypropylene and polyethylene to nylon to the higher performance thermoplastics like PEEK and PEKK. Typically, these are made by first making a continuous tape of fiber-reinforced thermoplastic, then this tape is chopped up and either pressed into a heated die for compression molding or formed into a biscuit of material that becomes a charge for a compression molding machine. Thermoplastic bulk molding compounds like this are used extensively in the auto industry now that most of the processing parameters have been worked out. What this means is that with thermoplastics, the temperature of the dies used to mold the product are higher than the softening temperature of the thermoset plastics and the temperature has to be held within a very narrow temperature band. This ensures that you not only completely fill the mold, which can be very detailed, but also that you also don't melt the plastic enough that the fibers wander away from the finer details of the part, which would make those details not a strong or stiff as they should be. Since the industries that are adopting this technology, notably the auto industry and commercial aviation, are extremely concerned about quality, repeatability, and reliability of their products—and rightly so—the manufacturers of these products have spent quite a bit of time and money perfecting their processes.

One interesting wrinkle in this field of thermoplastic BMCs is the use of longer fibers, especially carbon, in higher fiber fill densities and with the high-performance thermoplastics like PEEK and PEKK. These fibers are starting to move into the realm that would previously be addressed by either high-strength aluminum for very complex shapes or typical aerospace grade continuous fiber carbon/epoxy composites. This is because these higher performance thermoplastic BMCs with long-chopped carbon fiber have the same or higher strength and stiffness of aluminum at half the weight. Both the aerospace and the auto industries are constantly on the lookout for high-performance, lightweight materials that can be processed rapidly and inexpensively. With the focus on fuel economy in both fields, these materials are making inroads.

Metal Matrix Discontinuous Composites

Some of the first metal matrix composites (MMCs) were developed in the 1950s and 1960s during the height of the space race after the launch of Sputnik. Mostly these were continuous fiber MMCs with boron fiber in aluminum being one of the prevalent materials. At the same time, however, there was some experimentation with combining silicon carbide (SiC) with aluminum to make an MMC that could be forged and machined by some of the tools available at the time. These MMCs did not catch on until the early 1980s when the enhanced stiffness, strength, and ability to withstand high temperatures was needed for some space-qualified parts that got very hot on re-entry to the earth's atmosphere. These new "space-qualified" MMCs were primarily focused on SiC whiskers as a reinforcement because the whiskers act like short fibers and make the material stronger and stiffer.

Two different primary methods are used to make these materials. The first method that was developed is what is called *stir-casting*, where the particulate or whiskers are added to the molten metal and the molten mixture is stirred until the particles are distributed evenly. The other method is a powder metallurgy process where metal powders are mixed with hard particulates like SiC and the mix is put into a heated press under a vacuum until it becomes a consolidated material. Then to ensure that it becomes more like a metal than a sintered block, it is usually extruded at a high temperature, or it is subjected to what is called a hot isostatic press (HIP) that softens the metal because of the high temperature and consolidates the MMC. The resulting material can be forged, machined, and formed by nearly any metal forming process. Machining is slightly more difficult with the SiC reinforcement because normal machine tools, which are typically sintered carbides themselves, wear very rapidly so specialized diamond cutting tool inserts are typically used to do the machining on these materials.

The most common metals used for the matrix are aluminum and magnesium, although magnesium has somewhat fallen out of favor for general use because it has a tendency to corrode and is not as strong as its cousin aluminum. There is still ongoing work to develop alloys of magnesium that are stronger and less susceptible to corrosion while still being as light as the magnesium itself. Magnesium is about two thirds the density of aluminum, and a fifth the density of steel, so there is a push in several industries where weight is critical to enable magnesium MMCs to be applied.

Lots of aerospace and commercial products are made from aluminum-based discontinuous composites. Specialized Bicycle Components has been making bicycle frames out of the stuff for a couple decades now, and several auto makers have adopted aluminum MMCs for things like cylinder liners because of the material's strength, wear resistance, and ability to conduct heat away from the combustion chamber of their engines. This allows them to make lighter weight engines. Toyota, Honda, and Porsche have all developed and manufactured cars with aluminum MMC cylinder liners. Ford offers a boron carbide/aluminum MMC driveshaft for its performance cars that is much lighter than the aluminum driveshaft it replaces. Again, it is industries like automotive and aerospace that adopt these materials in efforts to make their cars and airplanes lighter, burn less fuel, and thereby reduce their carbon footprint. MMCs are also much more fire resistant, and they don't uptake water or outgas like plastic matrix composites do. They are also resistant to radiation damage, which is important for satellites and other spacecraft that are exposed to gamma radiation.

Non-Plastic Matrix Continuous Fiber Composites

I'm not going to dwell on these composites much in this book because they really are not mainstream materials. They are mostly relegated to those applications where their properties are required, and where cost is not that important. These materials are generally very difficult to make and therefore much more expensive than their plastic matrix cousins. There are three different matrices that are used to bind together the fibers in these composites: metals, ceramics, and carbon itself.

Continuous fiber MMCs can be made by diffusion bonding where thin sheets of the metal are placed between layers of continuous fiber, and the entire stack is hot pressed to consolidate it. In a sense then, these are much like continuous fiber plastic matrix composites, and the processing is similar to some thermoplastic matrix continuous composite materials. But since the temperatures and pressures to accomplish this diffusion are so high, these materials are difficult and expensive to make and typically are not made in large-volume manufacturing. Other methods of making these materials include pressure infiltration where the molten metal is pressed under high pressure into a bed of fibers, and also spray or plasma deposition where the molten metal is sprayed directly on the fibers.[1]

These continuous fiber MMCs are used primarily in aerospace, but some are used commercially as well. The F-16 uses a titanium matrix with silicon carbide fibers for part of the structure of its landing gear. Some high-performance sports cars use a continuous carbon fiber in a SiC matrix for disc brake rotors because of the high heat capacity and wear resistance of this material. For space applications, space shuttles had a boron/aluminum continuous fiber MMC strut in their landing gear as well as boron fiber/aluminum struts throughout their center fuselage section. The Hubble Space Telescope has a diffusion bonded graphite fiber/aluminum high-gain antenna boom. And, finally, graphite fiber/magnesium tubes were developed for DARPA[2] by Lockheed to demonstrate that these extremely light, stiff, and strong tubes could be manufactured reliably with high quality. But when carbon fiber composites came down in cost near the start of the present century, continuous fiber MMCs lost their main customer, the Department of Defense, and production and research very nearly came to a halt.

Ceramic matrix composites (CMCs) take a different tack on strength and stiffness than other composites. This is largely due to the fact that ceramics are already stiff and strong, they are just very brittle. Ceramic matrix composites try to solve that problem by incorporating fibers that have some toughness (ductility) into the otherwise brittle ceramic matrix. A great example of this is carbon fibers in a silicon carbide matrix. This material is as strong and stiff as silicon carbide, and in the fiber direction it has nearly 1 percent elongation to failure. What happens is that the fibers will bridge the small cracks that form in the ceramic and stop them from growing and causing a brittle failure of the ceramic. The most common mixtures used in ceramic matrix composites are carbon/carbon (yes, this extremely light stuff is characterized under ceramic matrix composites), carbon fibers in a silicon carbide matrix as mentioned above, silicon carbide fibers in a silicon carbide matrix, and alumina (Al_2O_3) fibers in an alumina matrix.[3] All but the alumina/alumina composites are made by depositing the matrix onto the fibers out of a gas mixture (called chemical vapor deposition), pyrolysis of a polymer matrix (this is how carbon/carbon composites are made), or chemical reaction to turn the matrix material from whatever precursor is used into a ceramic. This process is used to make CMC brake disks where a preform of carbon/carbon is made and then taken up to a temperature high enough to melt silicon, which diffuses into the matrix part of the carbon/carbon CMC and reacts with the pyrolyzed carbon to form silicon carbide. The other way to make CMCs is used for alumina/alumina and some other oxide-based ceramics where the alumina fibers are mixed with alumina powder in a preform and the result is fired in a furnace at the very high temperatures needed to fuse the alumina powder into a solid ceramic. Or, alternatively, using alumina fibers in a matrix material that consists of a silicate forming hydrocarbon (usually) and an alumina forming hydrocarbon to produce a mullite matrix $(3Al_2O_3 2SiO_2)$.

Well, I promised you I wasn't going to dwell on these composites as much as I have the traditional fiberglass and aerospace carbon/epoxy materials, so I'll stop here. Next up is a view of the composites business, where it came from, what it is today, and where I think it's headed in the future. And then I want to wrap up with a brief discussion of where you can go to get more education and training in composites and also where and how to get a job in this wonderful and fast-paced industry. There are great things happening in composites today and as they say, the world is your oyster—you never really know what you're going to get until you get involved and working in this business.

11

The Business of Composites

This is where it gets interesting. How do our branches interact with one another and compete for space in the canopy of our semantic tree? Or, how do the different companies compete with one another to come out on top and flourish in the sunlight? Sometimes it's just competition, and sometimes it's a pretty dirty business.

The business of composites and the history of that business is as full of backroom deals, "stolen" intellectual property, and people taking credit for things others have done, which is common in most other materials businesses. There have been lots of lawsuits over who first came up with this idea or that resin or this formulation. In addition to this, since the early days there has also been tremendous consolidation in the industry—especially in raw materials suppliers. This is more prevalent in the resin manufacturers (the glue people) than it is in the fiber business, but it is also true somewhat in the business of fabrication and manufacture of composites.

We went over some of this early on in the first chapter of the book, so it's time to expand on that a little bit, and also to provide you with information about where the composites industry is today, and where I think it is headed. The bottom line is that it has some of the feel of a maturing industry and some of the feel of a burgeoning new technology—it all depends on what part of the business you are looking at.

Basically, the business breaks down along two predictable lines: the raw materials providers and the fabricators of manufacturers of composites. Companies in each field take various approaches depending on their intended market. There are large and small companies in each piece or sector of the composites business. This is true more so in the fabricator side of the business than the material supplier side because of the scale needed to be a raw material supplier for a large industry. But there are still some smaller companies that produce and sell resins. West System, which is one of them, came out of the boat building industry.

String Supply Business

We have seen so far that glass fiber composites dominate the composites industry. This will be true for the foreseeable future as well. In this part of the book, I am going to touch on all the fibers and the business that surrounds each of them, and also their outlook for the future. The future of the fiber business is going to be concentrated on how to make these fibers cheaper, stronger, tougher, and more plentiful. Manufacturing technologies of today will be updated or, in some cases, completely replaced with newer, higher volume, less expensive processes. And the push to lower the precursor cost is also underway in the fiber business. This is especially true of the organic fibers such as aramids, polyethylene, etc. Let's start with glass fiber, which dominates the market.

Glass Fiber Business, Market, and Future

We learned in the first chapter of this book about the history of glass fiber manufacture and how it came about when a clever engineer noticed that long strands of glass were hanging from the ceiling in the glass bottle manufacturing facility where he was working. That company became the Owens Corning that makes all manner of things out of glass, including fiberglass insulation and high-performance glass fiber for the specialty composites industry. They also have a heavy investment in glass fiber for fiber optics, which is under increasing demand in the telecommunications and tech industry. Owens Corning is no longer the leading supplier of E-glass fiber in the United States According to *CompositesWorld*,[1] China is the leading glass fiber producer for E-glass worldwide. This is partly because of the low labor costs in China but also partly because glass fiber has become a mature commodity product. While most of the Chinese E-glass is not specifically geared for use in composites, they still provide the highest tonnage glass fiber for the commodity composites market— as in for SMC and BMC compounds. And a good chunk of that market is for domestic consumption in China. Some higher performance glass fibers are, however, now being produced in China. U.S. and European suppliers, while they still hold a good percentage of the market, have largely moved on to the more profitable and specialized applications and higher performance glass fibers (like E-CR, R, and S-glass), where the constant downward cost drive is less prevalent and where they can make a profit. But whatever is made and wherever it is done, there is a constant push for higher performance at lower cost. Both the auto industry and commercial aviation are helping make that push to increase performance and lower cost, so there is still a tremendous amount of manufacturing technology research and development going on even in the glass fiber business. The lower the price gets for a particular performance level, the more composites made from these fibers will replace metals in all sorts of manufacturing industries.

The future of glass fiber is very bright. While it suffered a bit during the pandemic, mostly because of supply chain issues, it is going to boom back and become an even larger business than it is today. According to *BusinessWire*,[2] the glass fiber business is expected to top $10.6 billion by 2026. A lot of this growth will be due to the burgeoning housing market where bathtubs, shower surrounds, and other interior finish products will see strong growth. The boat building industry, including recreational as well as commercial and military, will have increasing demand for glass fiber to build stronger, bigger, and faster boat hulls, as well as replace older boats as they wear out or fall apart. Other areas of growth for glass fiber will be replacing aging infrastructure made of steel, tanks of all varieties, printed circuit boards for all our tech and entertainment devices, wind turbine blades for the ever-expanding wind energy segment, and especially in automotive parts.

Carbon Fiber Business, Market, and Future

While glass fiber has been around the longest, and even though it's strength-to-weight ratio is not nearly as good as almost all other composite fibers, it still dominates the composites market, primarily because of its low cost. But that may be changing a bit in the not-too-distant future. As I noted earlier in the book, carbon fiber prices have been dropping dramatically for the last 20 years or so as fiber manufacturers have ramped up production, found new and more efficient ways to make high quality fiber, and lowered manufacturing costs throughout their supply chains. Toray of Japan is a case in point that I will get to in a couple of paragraphs.

There are a number of manufacturers of carbon fiber, and they are spread across the technologically developed part of the globe. Hexcel, first established in 1948 as an aluminum honeycomb company, is still the dominant U.S. manufacturer of PAN-based carbon fiber. They have manufacturing facilities in the United States and in Europe, and their primary market is aerospace, which is the dominant market for carbon fiber today. Hexcel has made the decision to stay in the higher performance carbon fiber business, specifically for the high-performance aerospace market, which is more profitable if lower volume, and has ceded the lower cost carbon fiber market to a few Japanese companies, Toray included. This fits with the company's origins as the first aluminum honeycomb company in the United States and nearly the lone supplier of honeycomb to NASA, the

DOD, and commercial aviation. Hexcel honeycomb was used for the landing pads on Apollo 11 and was also used extensively in the space shuttles. They have since moved into composite honeycomb but are still focused on the high-performance structural materials business for NASA and the DOD.

Another PAN-based carbon fiber manufacturer formed in the United States is Solvay Group. Originally called Cytec Engineered Materials, which spun off from American Cyanamid in 1993, it was acquired by the Belgian chemical and materials giant Solvay, S.A. in 2015. Cytec was a bit different than Hexcel and some of the other smaller carbon fiber manufacturers in that it had a completely vertically integrated supply chain, and already had an acrylic fiber business. This was because of their heritage as a part of one of the largest chemical and pharmaceutical companies in the world, American Cyanamid. But soon after the spin-off, Cytec also divested of its acrylic fiber business. Then, in 1997 Cytec acquired Fiberite, Inc., which was already in the advanced composites business although they did not actually manufacture carbon fiber. Since Cytec's business model was to be a vertically integrated company, they acquired BP Carbon, the major supplier of carbon fiber to Fiberite, in 2001. Originally part of BP Amoco, BP Carbon was the major supplier of carbon fiber to Fiberite, which at the time was part of Cytec Engineered Materials. This is a classic example of the consolidation that this industry has gone through over the years: as the industry has matured, manufacturers have consolidated or been acquired by larger conglomerates that focus more on particular parts of the business.

In Japan, there are several manufacturers of carbon fiber, mostly PAN fiber. This is partly because Dr. Akio Shindo of Japan's former Government Industrial Research Institute in Osaka invented the process for making carbon fiber from a PAN precursor in 1959, and the patent was owned by the Japanese Government. The U.S. and European companies initially made carbon fiber from rayon fiber both because they had established rayon businesses, and because PAN fiber wasn't used in textiles as much as other synthetics since it is difficult to process into fabrics. One of the oldest of these companies is Toho Tenax, which is now part of the Teijin Group. Toho Synthetic Fiber,[3] established in 1934 as a rayon fiber manufacturer, started making PAN acrylic fiber in 1963 primarily for the growing acrylic fiber textile business. Seeing that the future of carbon fiber would be promising, they started research into making carbon fiber in 1969, in the heyday of the burgeoning aerospace composites business. By the mid–1970s they could make adequate PAN precursors for high-performance carbon fibers, and in 1977 the manufacturing of these high-performance PAN precursors was initiated. By 1993 they had established Tenax fiber as their aerospace-grade carbon fiber and established manufacturing facilities in Germany and the United States—their first foray out of Japan. Then, in 2000 Toho became a wholly owned subsidiary of the Teijin Group and was renamed Toho Tenax. In 2004 Teijin acquired Fortafil Fibers (Knoxville, Tennessee) from AzkoNobel to strengthen their U.S. presence. Since that time, and with the advent of the Boeing 787 and Airbus A350 going into production, most carbon fiber manufacturers have upped their capacity to produce PAN-based carbon fiber. This is another example of consolidation and acquisitions in the composites business.

By far, the largest producer of PAN-based carbon fiber in the world today is Toray Industries in Japan. Along with their subsidiary Zoltec, they dominate the worldwide PAN-based carbon fiber market. Toray[4] started much as Toho did: by establishing a rayon fiber business in 1926, primarily for the growing synthetic fiber textile business that was looking for a silk-like fiber to make clothing that could be purchased by anyone. By 1941 they had moved into the nylon business by reverse engineering nylon that they acquired through Mitsui Bussan in New York. This was around the time that World War II broke out and Japan was very secretive about their industrial processes. After the war, DuPont sued Toray for infringement on DuPont's patents. Toray won that battle, but later set up an agreement with DuPont. With these successes under its belt, and after World War II when Japan decided to outcompete the rest of the world in technology and manufacturing, they established synthetic fiber R&D laboratories and started moving into production of other types of fiber, notably polyester fiber under license to Imperial Chemical Industries Ltd. in the UK in 1958. They continued branching out, producing acrylic and polyester fibers either under license or in partnership with U.S. companies (DuPont in particular). At the time of Toray's foray into PAN-based carbon fiber production, Union Carbide had already been making a version of the fiber based on Dr. Shindo's patent. Since they were already dominant in the textile fiber manufacturing industry in Japan and had already worked closely with other U.S. companies, the deal that Toray made with Union Carbide was extremely fruitful for Toray. Their original carbon fiber, TORAYCA™, was introduced in 1973 to a rapidly expanding aerospace market. Toray outpaced other carbon fiber manufacturers partly because of their already established manufacturing capacity and the cost reductions that ensued because of that capacity, and partly because their fiber was recognized as having high quality and performance at a lower cost. The quality and performance of the fibers is a direct result of the investment made by Toray and the Japanese government in their fiber research and development labs. Recognizing this, Boeing made Toray the major supplier of carbon fiber to its passenger aircraft. Since then, Toray has established manufacturing and R&D centers in the United States, Europe, and China, and has developed probably the widest range of PAN carbon fibers in the business today.

As for the future of carbon fiber, the automotive industry recognizes that they need to move away from steel if they are going to meet some of the carbon emissions targets that are looming. While there is little incentive to make the switch from steel to carbon—especially for things like the steel frame of a car—the challenge is to bring the cost of a carbon composite with at least 50 percent carbon fiber below $10 per pound.[5] Many in the industry have said that if that number could be reduced to as low as $5 per pound, carbon composites would immediately replace steel in the automotive industry. There is a long way to go to get there, but every year companies in this intensively competitive business either consolidate or increase their manufacturing capacity to lower cost. And with Toray still doing significant R&D in their manufacturing processes, the future may bring us all carbon fiber vehicles.

Aramid Fiber Business, Market, and Future

Aramid fibers came on the scene in the early 1960s with DuPont's Nomex. Since polyaramids were quite well known for their heat resistance, Nomex was marketed and still is marketed today as a replacement for asbestos. Nearly all the flame-resistant clothing in use by firefighters is made from Nomex. But since Nomex, which is a meta-aramid (remember the structure of the benzene ring, with ortho, meta, and para attachments depending on where they occur?) doesn't have either the strength or the stiffness needed to make a composite material, it has not been used in the composites business.

In the late 1960s to early 1970s, Monsanto and Bayer were doing research into aramids and developed a para-aramid fiber (attachments at 1 and 4 versus meta at 1 and 3) with much higher strength and stiffness. DuPont was watching this development because of their existing synthetic fiber business, so in 1973 with Stephanie Kwolek's invention of the fiber, DuPont introduced Kevlar on the market. Kevlar is still the dominant aramid fiber and is used extensively in the military as body armor, and in aerospace where it comes close to replacing the properties of carbon fiber.

And then AzkoNobel, who had been working with DuPont at the time, went on to produce a similar fiber, called it Twaron, with almost the same chemistry as Kevlar. DuPont immediately sued AzkoNobel for patent infringement, and the aramid fiber patent wars began. This eventually became the largest patent dispute of the century. Litigation went on for 11 years, during which time neither fiber did as well as it could have because customers were leery of settling on one or another fiber if they couldn't guarantee that they would be able to purchase the fiber in the long term. DuPont fared better than AzkoNobel did in the marketplace because they had been first to introduce para-aramid fiber to the world, and most U.S. companies figured that DuPont would ultimately win. In 1988, DuPont and AzkoNobel settled their differences and agreed to dismiss all lawsuits and market their fibers worldwide. Even though in the United States DuPont was winning, some court decisions favorable to AzkoNobel in the UK, Germany, and France[6] led Dupont to reconsider and drop its patent infringement efforts. Since that time, the Twaron fiber business was acquired from AzkoNobel by Teijin in 2000 and has become the Teijin Aramid Fiber business unit of Teijin Group.

Kevlar and Twaron fibers remain the only para-aramid fibers in the business, and both are used extensively in composites. There was one instance of a South Korean company, Kolon, stealing trade secrets from DuPont for the purpose of perfecting their own para-aramid fiber Heracron, but DuPont stopped that in 2015 when Kolon pleaded guilty to conspiring to steal trade secrets from DuPont. Kolon had recruited a former DuPont engineer and sales executive of by the name of Michael Mitchell, as well as a former DuPont engineer by the name of Edward Schultz. DuPont had suspected both Mitchell and Schultz of taking home proprietary confidential information about Kevlar, so the company alerted the FBI. When the FBI raided Mitchell's home, they found quite a bit of Kevlar trade secret information on his home computer. He went to prison for 18 months for that offense. It turned out that Kolon had been recruiting former DuPont employees since 2006 in an attempt to gain access to trade secrets and proprietary information. In 2009, the FBI charged several former Kolon executives who were in charge of the Heracron business unit with industrial espionage. None of them have traveled to the United States for trial, but the indictments still stand. In 2015, Kolon pleaded guilty to industrial espionage and agreed to pay $360 million in criminal fines and restitution.

As you can see, the composites business is replete with consolidation, backroom deals, patent infringements, and people getting convicted for stealing trade secrets from the bigger players in the industry. This is one story that has a good ending, but there are several where disputes are still raging, especially when former employees and executives of large material manufacturing companies leave to form their own companies and compete with their former employers.

DuPont has gone on to introduce two different grades or types of Kevlar, Kevlar 29 and Kevlar 49.[7] Each has its own niche and is used for different purposes, because the stiffness and ductility of these two fibers is rather different. Both grades of Kevlar have a tensile strength of around

8 ksi, which is slightly above both S-glass and carbon fiber and considerably above E-glass. Kevlar 29 has an elongation at break of 3.6 percent, which is comparable to E-glass and higher than carbon fiber, whereas Kevlar 49 only has an elongation to failure of 2.4 percent. Though the stiffnesses of these two aramid fibers are quite a bit different, the strength of each is about the same. The stiffness of Kevlar 29 is a little above 10 msi, whereas Kevlar 49 has a stiffness of over 16 msi. Comparable stiffnesses for glass fibers are 12.4 for S-glass and 10.5 for E-glass—about the same as Kevlar 29. Carbon fiber is much stiffer, with an elastic modulus around 30 msi or higher. Kevlar weighs a little more than half what glass fiber weighs, so its specific stiffness (stiffness/density) is nearly three times that of glass, which is why Kevlar is used where weight and stiffness matter.

Kevlar 29 is used mostly for bulletproof vests, helmets for soldiers, and things like the armor on Bradley Fighting Vehicles. There are other uses in the commercial and recreation industries as well where its higher strain to failure and very low density is more important than its stiffness. Climbing rope is a classic example of Kevlar use in recreation. It is also used in high-performance bicycle tires, bow strings for archery, ping pong paddles, tennis rackets, sails for high-performance boats, and even by Nike for running shoes.[8]

Kevlar 49 is the aramid fiber of choice for composites because of its very high stiffness-to-weight, or specific stiffness. In specific stiffness Kevlar 49 lies midway between carbon fiber and glass, so it finds its way into high-performance boat hulls as a replacement for glass fiber, and in aerospace applications where its light weight and high stiffness make it a good choice. NASA used it for pressure vessels on the space shuttle, and it has been used on the Trident missile for the solid rocket motor case. It has also been used in commercial aviation for fairings, like on the now retired Lockheed L-1011. Even Boeing uses some Kevlar 49 for fairings on several of its aircraft, including the 787, especially in areas like the engine nacelles and fairings over the engine mounts. Control surfaces on some aircraft are also made of Kevlar because of its high stiffness. It also makes a good mix with a carbon fiber part, especially as an outer layer because of its impact resistance.

The market and business in aramid fibers is only going to get stronger as DuPont and Teijin work on both the precursors to their aramid fiber products as well as their manufacturing processes. In fact, DuPont now has several more grades of Kevlar, each of which is intended for niche markets or applications. Research is continuing on the fabrication process for these fibers as both companies ramp up production to meet ongoing demand. This is actually pretty much the story of composites, especially in the material supplier side of the business—demand increases as the companies that make these products increase capacity and continue R&D focused on reducing cost.

Polyethylene and Polyester Fiber Business, Market, and Future

Ultra-high molecular weight polyethylene fiber is relatively new to the market, but it appears to be developing a strong impact in a number of areas, especially in the high-performance sailing yacht business. There are two competing fibers in this space, Spectra from Honeywell International in the United States and Dyneema from Dutch conglomerate DSM. These two fibers have an interesting history and there are good reasons why there are no patent disputes. They are both gel-spun ultra-high molecular weight polyethylene, also called high-modulus polyethylene, and both are made using the same process. As it turns out, a chemist at DSM made the original invention of this fiber accidentally by stirring a mixture in a beaker and noticing that fibrous tendrils were sticking to his glass stirring rod. In 1968, Dr. Albert Pennings[9] was working in the new polymer science lab at DSM, a Dutch company that had been able to transform coal into fertilizer but was now interested in polymer science. It was Pennings who noticed these little fibers sticking to his glass stirring rod, so he pulled them out of the solution, hung the small, fat single crystals of polyethylene above a Bunsen burner and stretched them out with his hands while the solvent was drying out of them. He knew that these were single crystals of polyethylene, so he recognized immediately that these would be very strong fibers because the longer the molecular chain or crystal, the stronger the plastic. As it turns out, he couldn't break them. Then he went running to his boss with his new fiber discovery and his boss yelled something like, "If I wanted fiber I would have picked up the phone and ordered some, now go back and make me some plastic." Undeterred, Pennings wrote several papers about polyethylene fiber, which gave him global fame and a professorship at the University of Groningen in Holland where he continued his research. It took until 1978 for DSM to ask one of Pennings' former PhD students, Paul Smith, to see if he could come up with a way to mass-produce Pennings' fiber. Paul worked with a colleague, also out of Pennings' lab, for about a year before they came up with the idea of spinning the fiber out of a gel rather than a solution. That was successful, and they patented the process in 1979. But, in 1982 DSM killed the project because it wasn't in their core business. It wasn't until 1990 that Dyneema first appeared on the market. But Honeywell had been working on making an ultra-high molecular weight polyethylene fiber as well, and at about the same time as Dyneema came out on the market, Honeywell had a patent granted for making what became Spectra.

Spectra fiber was originally developed by Allied Signal, whose researchers had read Pennings' patents. Originally conceived as a ballistic protection fiber, the first products that made it to market were things like Spectra Shield, which was marketed successfully as a replacement for Kevlar vests and handheld ballistic shields to both the military and law enforcement. It was also soon incorporated into ropes and slings, and has had tremendous commercial success, not only for the military and law enforcement, but also in the marine industry because it is nearly impervious to damage or degradation from salt water and is UV resistant. Since the two companies, DSM in Europe and Allied Signal (which bought and became Honeywell) in the United States, were initially both competing against Kevlar, they didn't engage in the patent disputes and other shenanigans that very nearly overcame DuPont. Both companies were probably wary of engaging in an intellectual property battle because they had seen DuPont and AzkoNobel's 11-year battle over Kevlar and Twaron, as well as the theft of trade secrets and people sent to jail.

Finally we get to Vectran, originally invented by Celanese Corporation and now manufactured by Kuraray in Japan. Vectran is the only melt-spun liquid crystal polyester fiber on the market. It was originally marketed in 1990, about the same time as Spectra and Dyneema. The fiber is golden in color and has very high thermal stability, low creep, and good chemical stability. In some applications, such as at very low temperatures, its thermal stability allows it to outperform aramid and polyethylene fibers. It does, however, tend to fray and has poor resistance to UV, so when it is used in things like marine rope or sailcloth it is coated with a UV-resistant coating like polyurethane or polyester. One of the notable applications of Vectran was the airbags that Mars Pathfinder used to cushion its landing. They worked well because of their inherent tensile strength while remaining very flexible in a fabric and maintaining their strength in the low temperatures of the Mars atmosphere. While there are several uses for Vectran, the total yearly production is still only around 1000 tons. To put this in perspective, Spectra fiber tonnage per year has expanded recently from 7500 tons to 8500 tons, and Kevlar production is more than 20,000 tons yearly. But since a rising tide floats all boats, the markets and futures for Spectra, Dyneema, and Vectran look very bright indeed.

Glue Supply Business

Since I need to cover the business side of the glues, rather than approach them from a historical perspective like I did earlier in the book, it makes more sense to introduce them in their order of usage or tons produced per year. I'm going to start with the highest tonnage, most prevalent glues and elaborate on the business of supplying these resins from a global market perspective. This leads us directly to the polyester and vinyl ester resin market.

As you will see coming up, the resin industry has had less backroom dealing and intellectual property theft than the string industry. There have been a few disputes along the way, but because the resin business came out of the oil and gas or coal by-product business—as did all plastics that have had commercial success—there has been more competition based on performance and less worry about patent protections. This is also due to the nature of the resins themselves. It doesn't take much work for a good polymer chemist to tweak an epoxy resin, for instance, and get a different formulation and also a different set of properties. So while there are patents in certain areas, in general there has been less tendency to sue and more tendency to improve on someone else's formulation and make something completely new and "better" than the other guy's resin. Fiber production and fiber processing is far harder to tweak in a beaker.

Polyester and Vinyl Ester Resin Business, Market, and Future

There are a number of major players in the unsaturated polyester resin business today, and all are very large conglomerates with a broad range of specialty chemicals. Some of the names you have heard, like DSM of Dyneema fame. Others headquartered either in the United States or Western Europe are Ashland Specialty Chemicals, headquartered in Illinois; Dow Chemical, headquartered in Michigan; BASF, headquartered in Germany with a large presence in the United States and U.S. headquarters in New Jersey; and Polynt, another European conglomerate headquartered in Italy. One of the first and oldest companies in the unsaturated polyester business is Reichold. It was initially founded by a German immigrant in the United States, Henry Reichold, who was working for Henry Ford in the paint department. Ford was looking for a paint that would dry quickly to replace the lacquers and enamels that took days or weeks to dry. Henry Reichold used some of the chemistry and resin that was being made in Europe by his family's business and formulated a paint that would dry in minutes. That was the start of Reichold. There are a number of large chemical conglomerates in Asia as well, and a number of those companies have acquired either U.S. or European companies. Production of these resins is accomplished by these large conglomerates in factories all over the world.

In 2021, the total market for unsaturated polyester resin was a little over $10 billion, with a projected growth rate of more than 6 percent per year. One study[10] projects that the market will be as large as $17 billion by 2027. This is largely because of the expansion of the construction industry, where cast polyester resin countertops, shower stalls, bathtubs, etc., are in high demand and have replaced more traditional materials. This growth is also due to the expansion of the fiberglass composites business as more and more consumer products, auto body panels, aircraft parts, recreational and also military boats, transition from traditional metals to fiberglass composites. In addition, the wind energy business is booming and will continue to boom as we work to get humanity off the hydrocarbon economy and become carbon neutral. Wind turbine blades are almost entirely composite and mostly glass/polyester or glass/vinyl ester, although the vinyl ester resins are somewhat more expensive and are used mostly where cost is not as important. Some of the larger wind turbine blades use quite a bit of resin; they are typically about 50 percent resin by volume.

Some of the same major players in the polyester resin business are also in the vinyl ester resin business. Most notable of these is Polynt. They merged with Reichold in the United States in 2017 and now account for a large proportion of the production of vinyl ester resins, especially those made in the United States. Polynt has a history of acquisitions and buyouts very familiar to the composites industry. Polynt was originally founded in Italy in 1955 as a producer of phthalic anhydride, which is one of the precursors to polyester and vinyl ester resins as well as a number of other plastics. Over the years the original company, which was called FTALITAL in the beginning, either acquired other companies or was acquired. Along the way their name changed many times, as did their corporate culture and philosophy. Now they are a completely vertically integrated composites raw material supplier as well as a large composites fabricator.

Another rather well-known name in the vinyl ester business is INEOS. This company is a global conglomerate organized around several businesses, one of which is polymers and composites. INEOS is another acquisition and conglomeration story. The company was formed in 1998 by two Brits named Sir Jim Ratcliffe[11] and John Hollowood, who had purchased the chemical arm of BP in 1992 to form Inspec. Then in 1995 Inspec bought BP's ethylene oxide and glycol business, and in 1998 Ratcliffe formed INEOS to purchase the ethylene oxide facility in Antwerp that Inspec had acquired from BP. Since then, INEOS has grown through the acquisition of major chemical business giants like BP, ICI, Amoco, Dow Chemical, Solvay, and BASF, into one of Europe's largest corporate conglomerates, providing both raw materials to the plastics and composites industry, as well as finished products in those industries.

One more large player in the vinyl ester resin business is AOC, headquartered in Collierville, Tennessee. Formed in 1961 from a private equity buyout of Owens Corning's chemical business, AOC has been bought and sold several times, moved their headquarters to the Netherlands, and as recently as 2021, was purchased by Lone Star Capital. They have specialty resin and chemicals plants in the United States (Owens Corning), Canada, Mexico, Europe and Asia.

The Asian market for vinyl ester is dominated by Sino Polymer with their MFE® brand. Formed in 1970 by Professor Zhou Runpai from East China University, it is the current state-owned polymer resin R&D and manufacturing company. Though they originally made epoxies, over the years Sino Polymer branched out to phenolics, high-performance polyesters, and vinyl esters. During the early 2000s they branched out again and started to engage in overseas markets with their resins as China began competing on the global marketplace. Since then, they have established manufacturing facilities in France, Saudi Arabia, and Malaysia. They claim on their website that they produce 20,000 tons a year of vinyl ester resin for commercial and industrial markets.[12]

The total market for vinyl ester resins is much smaller than that for polyesters, valued in 2020 at a little under $1 billion, which slightly declined from 2019 because of the pandemic. It is expected to grow by a little less than 5 percent per year for the foreseeable future.[13] This resin system does has a bright future because costs for manufacture have been decreasing steadily, so vinyl ester resins will begin to displace some of the applications for polyester resins because of its inherent corrosion resistance.

Epoxy Resin Business, Market, and Future

The epoxy business is the second largest resin manufacturing business in the world, after the polyester resin business. There are several major players in this business, such as Dow Chemical, BASF, and Solvay. However, there are a few others that specialize in epoxy resins only. Hexion Specialty Chemicals is one of the larger of these, formed out of a merger in 2005 between Borden Chemicals, Resolution Performance Products, Resolution Specialty Materials, and Bakelite AG of phenolic fame.[14] And then in 2010, Hexion Specialty Chemicals merged with Momentum Performance Materials to form Momentive Performance Material Holdings LLC, which in 2015, renamed itself Hexion Inc. They are still headquartered in Columbus, Ohio, and have a couple of different brands of epoxy resins they manufacture. The most common of these is the Epon® series of resins, which is quite well known in the aerospace industry and was specified for high-performance carbon composites. Prepregs of Toray's T800 and T1000 are commonly made using Epon 828 BPA-based epoxy. This is in fact one of the first of the high-performance aerospace epoxies that came on the market in the early days of high-performance composites, which is why it has become something of a standard in that industry.

Huntsman[15] is another major epoxy and urethane resin manufacturer and also has an interesting story. John Huntsman formed Huntsman Container Corp. in Fullerton, CA, and one of his first products was the clamshell Big Mac container for McDonalds. John went on to form Huntsman Chemical Company in Salt Lake City, Utah, in 1982, and bought the polypropylene plant in Woodbury, New Jersey, from Shell Chemical when Shell started shedding its specialty chemicals business. In the 1980s and 1990s, Huntsman went on to purchase the styrene manufacturing facilities from Celanese, the plastic film and flexible packaging business from Goodyear Tire and Rubber, the alkyl benzene and maleic anhydride business from Monsanto, and in 1994, the worldwide operations of the Texaco Chemical Company as well as the polypropylene business of Eastman Chemical Company. Finally in 2004, after several more acquisitions and few divestitures, Huntsman moved its corporate headquarters to The Woodlands, Texas, just north of Houston. By then Huntsman had acquired ICI's polyurethane business, which nearly doubled the size of Huntsman. That acquisition made Huntsman the third largest petrochemicals business in the United States. Their advanced materials offerings are mostly urethanes, epoxies, acrylics, and high-performance adhesives for the aerospace industry, automotive (BMW in particular), and electrical power transmission industries. They are one of the largest polyurethane foam manufacturers in the building insulation business in the United States. Their epoxy resin product is called ARALDITE and there are a range of epoxies with different additives for various applications. They are even introducing a new conductive epoxy adhesive with carbon nanotubes as the conductive medium to embed into composite structures as heat generators.

One surprising major epoxy manufacturing company—at least to those who have been around a while and know this company—is Olin Corporation. Until 2015, Olin was a gunpowder and cartridge company and owned the Winchester brand. While Olin has gone through some good times as well as bad and has bought and sold quite a few businesses, in the early 2000s they refocused completely on the ammunition business. That business remained very strong throughout the Afghanistan conflict because one of Olin's primary customers is the U.S. Department of Defense. In 2015 the Dow Chemical Company announced that it was going to sell its epoxy and chlorinated organics business, and Olin acquired the business completely, making it one of the largest epoxy manufacturers in the country. They call themselves the premier global manufacturer of epoxy products, but this is mostly just marketing hype. They do have the full range of epoxy products that Dow had, including liquid laminating resins, epoxy Novolac resins, epoxy adhesives, water-based systems, solid epoxies, and even powders for the epoxy powder coating business. And since both the epoxy and ammunition businesses are booming right now, Olin is slated to do very well into the foreseeable future.

The Asian market in epoxies is comprised primarily of three companies, Kukdo Chemical in South Korea, and Chang Chun Group and Nan Ya Plastics in Taipei. While Chang Chun is the largest of the three and dominates the Chinese market, the other two have significant market share and a broader reach globally than Chang Chun, especially in the special applications epoxy markets. And Nan Ya is the only one that has committed to a carbon-neutral future.

One more to note is Atul in Gujarat, India. Formed immediately after the independence of India from British rule in 1947, Atul has become one of the largest specialty chemicals companies in India. They have subsidiaries in the United States, UK, China, Brazil, and the UAE. They pride themselves in their R&D, inclusive corporate culture, and spirit of continual learning and innovation. Their businesses range from supply of aromatics, mostly from plant extracts, bulk chemicals and intermediates, to botanicals, pharmaceuticals, and high-performance and retail polymers. By high-performance polymers they mean epoxies; Atul is the largest manufacturer of BPA-based epoxies in India.

The global epoxy business in 2020 was valued at a little over $11.6 billion and is expected to grow at a compound annual rate of 6.7 percent[16] to nearly $20 billion by 2028. Quite a bit of this growth is expected to come from the Asia-Pacific region of the world with China and India leading that market. This growth is attributed to both the housing and automotive industries and is primarily in epoxy paints and coatings. The global epoxy composites market was on the order of $20 billion in 2020 and is expected to grow to a little over $47 billion globally by 2028.[17] This bodes well for the laminating epoxy market, all the epoxy resin suppliers, as well as the fiber and textile fiber suppliers and fabricators of epoxy-based composites.

Phenolic Resin Business, Market, and Future

Finally, we get to the phenolics market, the one started by Leo Baekeland when he first mixed his resin with finely divided paper and wood flour to create Bakelite. Since that time, other resins have found more use for composites primarily because of the higher fabrication cost of using phenolics over other, easier-to-use resins, the caustic chemistry of the cross-link reaction (sulfuric acid and formaldehyde), the high temperatures required to cure them, and the fact that they outgas water vapor during the cure process, which makes little bubbles come to the surface and mars the finish. They are, however, more flame retardant and fire resistant than polyesters and most epoxies and are very heat tolerant, so they find use in areas where there will be higher temperatures. They also get used for electrical insulators because, along with being very heat tolerant, they are stronger and harder than a lot of other plastics. For composites, they are used in applications where high heat resistance is needed, like on leading edges of supersonic aircraft, or re-entry shields for spacecraft.

The resole phenolic resins are used primarily in the wood products and related industries, and as curable adhesives. Oriented strand board (OSB), which has become the sheathing material of choice in the construction industry, is basically wood chips stuck together with a resole phenolic and heat pressed into a board. Also, all the laminated beams that you see in new construction are layers of wood glued together with phenolic resins.

In the composites industry, it is mostly Novolac phenolics that are used, although there are some composites that use resole resins. The phenolic composites industry was essentially in a hiatus with very few applications, mostly household appliances and some industrial products, which were copies or offshoots of Bakelite. That was until the Space Race started and NASA needed to invent a material that would not catch fire on re-entry but had very high heat tolerance, good insulating properties, and could ablate rather than melt when coming down through the upper atmosphere. Phenolic composite heat shields were used on the Mercury, Gemini, and Apollo space capsules, and are also used on the SpaceX Crew Dragon that has shown considerable success of late.

NASA also needed materials for rocket nozzles that could survive the heat of the rocket blast but were lighter than metals or ceramics. The original Saturn V engines designed by Werner Von Braun used Inconel tubes wrapped in a conical fashion and welded together so that the kerosene fuel would run through to cool the nozzle enough that the Inconel would not melt. But for solid rocket boosters and missiles, this sort of design wouldn't work because there was no liquid to flow through the nozzle, so solid rocket boosters have either a carbon-carbon composite nozzle or, more frequently, a phenolic-carbon composite with carbon black added as a thermal absorber and stabilizer. The STS solid boosters had a nozzle built up of layers of carbon fiber/phenolic resin that was made in the form of a tape, which was wound into the interior surface of the nozzle by a winding machine.[18] This machine would blow 700-degree air onto the surface to soften the phenolic and a roller would press it into place under high pressure to eliminate all the air. Then the consolidated tape would be given a blast of -40-degree CO_2 to freeze the phenolic to prevent it from curing as the next layer was laid down over it. Other areas of the nozzle less prone to exposure to the hot rocket exhaust used a glass/phenolic tape as an insulating layer to keep the metal structure of the nozzle from melting or softening.

Major players in the phenolic resin market are Bakelite, Hexion, BASF, Kolon Industries, Chang Chung Group, and Mitsui Chemicals. There are plenty of other smaller players that work in niche markets, but these are some of the largest.

One notable company that I have not yet mentioned is Plenco. Initially formed in 1934[19] as American Molded Products Company in Chicago, they started using phenolics to cast molded products like handles, knobs, radio cabinets, and other mostly consumer products, as well as trim parts for the automotive industry. They moved to Sheboygan, Wisconsin, and changed their name to Plastics Engineering Company (so you see where the name Plenco comes from). In 1939, they switched from liquid phenolics to powdered resins that they could mold into these same parts and stopped making liquid resins for some time. During World War II they supplied numerous knobs, handles, etc., as finished products for the war effort. They thrived during the war since the Army needed quite a bit of their material. But, unfortunately, after the war the supply of powdered phenolics dried up so the company had to switch back to making their own liquid resins, and they chose a phenol-formaldehyde resin formulation—i.e., a phenolic. They make both resole and Novolac types and sell molding compounds, adhesives, and laminating resins. They are now one of the largest U.S.-based phenolic resin manufacturers. They have maintained Plenco as a family-owned business throughout the years and have not been acquired, nor have they acquired many other companies. The three exceptions to this are when they acquired GE's phenolic business in 1982, when they acquired the Valite brand of phenolic molding compounds from Valentine Sugars, and when they acquired the Plasloc brand of phenolic molding compounds and the PlasGlas polyester bulk molding compounds from Plasloc Corp in 2020.

The global phenolic market was about $11.7 billion in 2021,[20] attributed in large part to the molded wood products industry using tons of resole resins for OSB, laminated beams, and adhesives. It is projected to reach $14.4 billion by 2026 and will grow at that rate for some time. There is some hope for faster growth, however, as the automotive industry transitions to lighter materials for primary parts both for internal combustion engines as well as housings and other parts for the electric vehicle future. Phenolics can replace metal parts on engines readily because of their very high tolerance for heat and their inherent strength and stiffness. Watch to see how many auto makers adopt more phenolic parts for their internal combustion engines as well as things like the electric transaxles and their housings.

Composite Design and Fabrication Business

The composites fabrication business is enormous and growing. Most of the major aerospace companies do their own thing and have their own composite shops. And the companies that fabricate composites for all industries run the gamut from small family-owned businesses to enormous corporate conglomerates. According to Composites World's "Top Shops,"[21] the 10 top performing composites fabrication companies worldwide are the following:

- Arisawa Mfg. Co. Ltd. (Joetsu, Japan)
- B&T Composites (Florina, Greece)
- Bucci Composites (Faenza, Italy)
- Cecence Ltd. (Salisbury, UK)
- Champion Fiberglass Inc. (Spring, Texas, U.S.)
- EPP Composites (Rajkot, India)
- GSE Dynamics Inc. (Hauppauge, N.Y., U.S.)
- The Gund Co. (St. Louis, Mo., U.S.)
- Oribi Manufacturing (Commerce City, Colo., U.S.)
- Prathamesh Industries (Ahmednagar, India)

These companies are in several fast-growing industries in composites, such as automotive, construction and infrastructure, the marine industry, mass transit, medical devices, oil and gas, and wind energy. They are less involved in the aerospace and defense indsutried, but still have a solid presence in both. That same article has a table of the industries served by composites fabrication companies, which is shown below.

Industries Served by Top Composites Shops (from Composites World)

Industries Served	Top Shops	Other Shops
Aerospace, Commercial	41%	52%
Aerospace, General Aviation	31%	39%
Agriculture	13%	12%
Automotive	33%	36%
Construction/Infrastructure	44%	24%

Industries Served	Top Shops	Other Shops
Consumer	18%	23%
Defense/Military	38%	47%
Industrial/Corrosion Resistance	28%	24%
Marine	41%	29%
Mass Transit	23%	15%
Medical	33%	19%
Oil and Gas	28%	13%
Sports and Recreation	23%	31%
Wind Energy	26%	18%

The total composites market in 2021 was $88 billion, and is projected to grow to about $126 billion by 2026.[22] This is due to a number of industries that are in major growth spurts, and also industries that are making the transition from metals to composites. Wind energy is going to be very strong and will grow substantially for the foreseeable future, with offshore wind farms being planned and permitted all over the globe in coastal waters. These make tremendous sense, because the population that needs the electricity these offshore wind farms will generate lives largely along coastlines where arable land exists because of the influence of the local oceans. The automotive industry has been making the switch to composites to replace metals for about the last 10 years at a steady pace, but with the new fuel economy standards kicking in soon, that pace will only accelerate. The transition to electric vehicles will also force the auto industry to move more primary structure to composites.

Other factors driving the composites market are government stimulus packages. China has been using this instrument almost like a weapon for several years, investing heavily in what they perceive as the industries of the future, including composites. Another example in the United States is the recent passage of the stimulus package and its focus on small businesses, infrastructure, health care, and other industries that are already making the move to composites.

In terms of demand and volume, however, the transportation business and the automotive industry lead all others. The transportation industry includes not only aviation, but also mass transit, rail travel, even the upcoming high-speed transportation systems like Elon Musk's Hyperloop. All these are going to need to move to composites to lower weight and increase efficiency. The transportation industry needs to lower their cost per person-mile or for goods per ton-mile.

The key drivers in the industry for improvements are the materials and fabrication costs of traditional glass fiber and carbon fiber thermoset composites. There has been an enormous amount of work done to reduce costs of layups through automation, like automated tape laying machines. There has also been a great effort to lower the cost of the fibers and resins themselves. As I mentioned earlier, it was once said that as soon as carbon/epoxy composites came down in price to $5 a pound, the automotive industry would stop using steel altogether. That number is somewhat fictitious, and in today's dollar terms probably doesn't apply. But the sentiment is there nonetheless. There is a price point below which automakers will gladly make the switch. And, for large parts of today's automobile, there are composite parts and panels that would have been unthinkable or unaffordable just 10 years ago. The bottom line is that composites have a very bright future indeed, and at some point will become the material of choice for primary structure in several industries.

12

Jobs and Schools
in Composites

Finally, let's see if we can get our semantic tree to bear fruit. What I mean by this is getting a job in this business and/or getting an education that will lead you to a good job. Believe me, once you become successful on this path you will be able to put food on the table for the rest of your life.

You don't have to be an engineer to get a job in composites, but it does help quite a bit. And if you're going into the field wanting to design things made of composites, it really helps to have a degree in mechanical engineering (ME) from one of the universities that has a good composites program. Later in this chapter I provide a list of those schools and detail what part of the composites business they specialize in. If you do decide to go that route—to get a degree in mechanical engineering—it is wise to get at least a master's degree, and to do something in composites for your master's thesis or project, whichever you think is appropriate. This will also give you a chance to do a work-study or an internship at one of the composites companies that was outlined in the previous chapter. That experience is worth more than just getting the degree because it's the hands-on learning experience that is critical to success in this industry. It will also give you exposure to a company you might be able to work for when you get out of school.

One thing that won't work for you is to look on sites like Monster and Indeed because these are really just job boards. They won't give you the information you need to get a job in composites. Indirectly, however, they can be a great source of information for research to find companies that are involved in composites. You can select those that you want to target. The rest will be up to you and how much you really want this. It takes persistence and patience, and you may have to knock on lots of doors before you find what you want.

Getting a Job in Composites Without a BS Degree in Mechanical Engineering

If you don't want to or can't afford to go to college, there are other opportunities for you. One of the best ways of doing this is to look locally to see if there is a company that fabricates things out of composites. There are, as we saw in the previous chapter, a lot of companies both large and small that make things out of composite materials. Looking online for composites shops in your local area is a place to start.

Another way to find out which nearby composites companies might be looking for people is to go to your local community college or possibly a trade school and ask if they work with any companies that deal in composites. Most of the composites companies pick from local trade schools to find technicians who are interested and have been trained, hopefully in a lab setting, how to make things out of composites, and also how to read a drawing or process document. Most companies have good process documentation for how they make molds, do layups, whether they use prepregs or do infusion molding, how to cure the part, and how to inspect it when it is finished. In fact, if you can afford it, attending a 2-year course of study in composites at a local community college is a great way to get the kind of experience and training that the better composites shops are looking for.

Finally, the way some have done it is to just start messing around with the materials. Start with a design that you find online, get some string and some glue, build a mold, do a layup and infuse it with resin, cure it, and take it out of the mold to trim off the rough edges to make a useful part. It doesn't have to be anything as large as a 30' sailboat, but a scaled-down model of something would work. That's how amateur boat builders become people like the Gudgeon Brothers of West System epoxy fame, or Bill Seemann of Seeman Composites who invented the SCRIMP vacuum infusion process. Quite a few of the smaller composites shops you will find started this way, so it is possible to start your own business once you get a little experience. Of course, you have to find a market for what you make, so it pays to start small.

Getting a Job in Composites with a BS in Mechanical Engineering

The decision whether to pursue going to college to get a degree in mechanical engineering is sometimes a difficult one to make. If you are already on that and have a plan to get there, that's half the battle—and that's great, by the way. It's a wonderful training ground for the rest of your life because you will always be able to find a job somewhere. Even companies like Apple and Microsoft have lots of mechanical engineers on staff, especially if they make computer hardware or any sort of physical device. Mechanical engineers learn quite a bit about how to program a computer during their education.

I can speak from experience here. I converted a FORTRAN program that calculated the power coefficient of a horizontal axis wind turbine, which is the first thing you do when designing a wind turbine and its blades, to run on an HP-41C calculator. You may not know what that is, but in the heyday of HP calculators, there was one that had not only all the math functions on it, but it also had a decent programming language. It was sort of the first high-performance programmable calculator on the market. My professor Bob Wilson of Oregon State had been working in the wind industry when DOE was just starting to fund research into development of the large wind turbines like those you see in the hills near Livermore, California. This was at the height of university research into wind power machines, and one of the hubs was Oregon State. That experience has served me for my entire career, because quite a bit of it has been in the field of structural mechanics and materials—especially composites. Hence this book.

On the other hand, if you have not yet decided whether to pursue a college degree, and especially one in ME, there are a number of things that you need to consider. First, are you any good at math and physics, or does it just leave you confused and frustrated? If so, engineering is probably not the place for you. You can make it through, but you will have to study 5 times as much just to make average grades. If, on the other hand, you at least understand the concepts but need a little more practice and training, or if you are an average or above-average student and you have others around you who can help you learn this stuff, then the decision gets a little easier. Also, if you have a limited budget for school, most state schools offer lower tuition to in-state residents, and if you're lucky you can get some work study through the university, or a job at a local composites fabrication shop. This road is a good one if you want to make a career out of composites, because if you are careful about what school you pick, or if your state school is in a city with some composites design and fabrication shops, you might be able to get some hands-on experience while you're going to school. That will jump start your career in this fast-growing industry because as the industry grows and the old guard (like myself) start to retire out of it, it needs new, young, experienced people to take over when they leave.

In any case, the decision whether to pursue a BS degree in ME rests on several factors. First is how well you grasp the fundamentals of the math, physics, and materials that make up the core of the composites design and analysis business. If you understand what I've presented in this book—at least at the basic level, if not some of the more advanced material—you probably have the learning skills to be able to tackle getting a degree in ME. The second part of this decision always has to do with the availability of funding to get you through college. Remember, you will need a place to live and money for food; you can start out in a dorm, which is a fixed cost per quarter, or semester, or year in some cases. The third part, and this is a little more difficult to discern, is whether the university that you have chosen is connected somehow to the composites industry. Do they participate in any of the internship programs at major aerospace and commercial companies that are looking for young talent? Are any of the professors in ME working directly with DOE or DOD, or some of the larger composite materials suppliers or manufacturers? This is where you will have to do some digging to figure out how you are going to go about getting on the radar of either the professors or one of the companies that offers internships. Internships from good engineering schools are a ripe source of training and experience, so you must pick wisely. If you want to be a composites engineer, you have to go after the internships that are offered by the companies you have decided to target because they work in composites engineering and that's what you want to do.

Now that you have all those decisions made and you have done enough research that it is starting to look like you might be able to pull this off, you need to start thinking about electives and what's offered. You are looking for the elective courses that will put you on the path to a career in the composites industry. This is especially true in the aerospace and automotive industries. These composites companies don't necessarily need to be the Big Three or one of the major aerospace companies like Boeing, Northrop, or Lockheed, although that is a good place to start. You're looking for what are called first-tier suppliers to those industries. These are the companies that make the ailerons, door panels, trunk liners, aircraft control surfaces, wing skins, you get the drift. Most of the major companies are more designers and assemblers than they are completely vertically integrated. This is certainly true of the aerospace business, where major pieces of aircraft are built by first-tier suppliers and then assembled by the prime aerospace company. Quite a few of the first-tier suppliers are large enough that they have internships themselves because they are looking for talent coming from the universities.

Now that you have a plan for how to get a BS in ME, and how to get a job as a composites engineer when you graduate, it is up to you to carry out your plan. Two words of advice that I give all budding engineers. First, you are young, and you probably have a boyfriend or girlfriend, and you may be thinking of getting married, or maybe you already are. Having kids as soon as you get out of school might be a mistake, unless you have already gotten an advanced degree. If you have an advanced degree and you think you might want to get another one, it is still a good idea to wait on having kids. It's a little easier on the guys than it is on the gals, because most of the child rearing is usually done by the woman. In my case, we had to share those duties because my wife got a degree in pharmacy and got a job in a hospital right out of school. When the kids came, she had to be at work. I had a bit more flexibility in the R&D business, so I ended up dropping the kids off and picking them up from daycare, helping with their homework, the cooking, and at least half, if not more, of the cleaning. No easy feat when you're trying to start a career, believe me. This is what I mean when I say I have the tire tracks on my back to prove what not to do.

Which brings me to the second piece of advice I give every budding engineer. If possible, get at least a master's degree in either ME or aerospace, or even composites engineering if your school offers it. That is, if you feel that you are talented enough and want to learn more and make quite a bit more money. If you're to this point in this book and you have that instinct in your gut that says that you want to and can do this, then by all means go do it. It would please me to no end to have influenced at least one or two people to get into this business and make a career out of it.

Good Composites Universities

This is not intended to be an exhaustive list of universities in the United States that have good composites programs. It is merely a selection of those that I am very familiar with and that have a good reputation in both industry and academia in composites. At the top of the list of course is where I completed my doctorate, the University of Delaware's Center for Composite Materials (CCM). This list is not, however, just about my favorites. Rather, it is a list of those universities that have a long history of training composites engineers and that have not only good programs, but also strong ties to industry. Most of these are hands-on institutions that teach the theory and practice of composites design and analysis. Some of them work directly with industry partners (like Delaware), and most of the better ones do consulting or research directed at helping these industrial partners either solve problems, refine a manufacturing process, or even develop a new manufacturing process that will give them an edge in the competitive business of composites.

Some of these programs focus more on design/analysis, whereas others focus more on manufacturing. All the programs are great at both and have very active relationships with the industry and with government experts in composites at all the major government labs. This is especially true of the DOD labs, since that is where most of the research, design, manufacturing and application of composites is happening in the government.

University of Delaware's Center
for Composite Materials

This campus is in Newark, Delaware, 25 miles from Wilmington, which is the home of DuPont. The University of Delaware has had an ongoing relationship with DuPont for decades. But, more importantly, they have what they call their University Industry Research Consortium,[1] which is a research and development program funded by the companies that join the consortium. While the consortium used to be a pool of money that was used to guide research projects that all companies were interested in, now it has morphed into more of a means for a company to find a particular research project they are interested in and guide the research in the direction they want it to go. Essentially it is more a university-company partnership for each project. This is one of the best and longest lasting collaborative research and development programs in composites the United States, and it is extremely successful.

If you go to the CCM site I referenced in the note above, at the bottom of the front page of that site is a link to a listing of "CCM and Industrial Partner Jobs." This was initially part of the motivation for creating the center, and for forming an industrial consortium. The center is very closely tied to both the ME and chemical engineering departments and gives undergrads the opportunity to work on industrial research and development projects while getting their BS degrees. They also offer MS and PhD programs through CCM itself, so this has been one of the best training grounds in the United States for graduate-level engineers to get jobs in the composites business. And, of course, being a short hop to Wilmington and DuPont doesn't hurt.

Northwestern University

At Northwestern, the composites studies and research are conducted in the Materials Science and Engineering Department.[2] They focus mostly on the nano-scale and micro-scale properties through design and processing of these materials to achieve desired properties by understanding the microstructure of the materials. They have world-class characterization facilities and focus primarily on understanding and explaining the properties of composites from the microstructure level.

Their industrial collaboration is not nearly as extensive as the University of Delaware's, but they do have several members of their advisory board who are senior figures at companies like Dow, BP, and Johnson & Johnson, as well as some of the larger government laboratories. Their curriculum is more focused on the material science aspects of composites than it is on the design/analysis/fabrication side of the business, so their graduates work in labs or for companies like Dow and DuPont that study and characterize these materials in their own laboratories.

MIT

MIT has what they call the necstlab,[3] pronounced "NextLab." This lab is part of the Aeronautics and Astronautics Department at MIT, which is one of the oldest aeronautics departments in the United States. Their focus is primarily on the field of aerospace and they do fundamental research in the aerospace composites business. They do have an extensive amount of cooperation with the Mechanical Engineering, Materials Science and Engineering and Chemical Engineering Departments. They focus on some of the newer advances in composite materials like carbon nanotubes, nanoengineered electrical actuators, polymer nanocomposite mechanics, and nanoengineered hybrid composite material architectures to enhance aerospace laminate performance.

They also have an industrial consortium initiated in 2006 with the likes of Airbus, ANSYS, Embraer, Lockheed, Saab AB (the aerospace side of Saab), SAERTEX, and Teijin Carbon America. This is another university-industry collaboration that is bearing fruit and provides graduates and students opportunities to land a career-level job in one of these companies. And, of course, MIT has such a good reputation that nearly anyone would be willing to hire one of their graduates.

Georgia Tech

Research and faculty in composites at Georgia Tech are again headquartered in the Material Science and Engineering Department,[4] just like at Northwestern. Their work mostly encompasses micromechanics of processing and fabrication of composites, which is where a lot of the work must go to make these materials better as well as less expensive. And their composites work is interdisciplinary which is necessary to teach undergraduates and graduate students what they need to know to get good jobs in companies focused on making composites better performing and more affordable.

Georgia Tech also has a corporate engagement group that solicits research projects and grants from corporate sponsors. The group offers companies that pay into it access to student resumes as well as space for one-on-one interviews with the students. They also offer a mentorship program wherein one of the corporate senior technical folks will mentor a student or two as they go through their studies. This is another great way for these companies to identify talent and obtain qualified young composites engineers.

Stanford

Stanford is as legendary as MIT, and their composites work is of the same caliber. Like MIT, their Structures and Composites Laboratory[5] is part of the Aeronautics and Astronautics Department in the School of Engineering. Their focus is on multifunctional structures, intelligent structures, stretchable sensor networks, and state estimation and structural health monitoring. Their research spans from bio-mimetic materials to structurally integrated lithium-ion batteries (rather than the battery trays of today's EVs). The lab faculty have been in the composites research and development arena for most of their careers and are now emeritus professors. This includes Dick Christensen, George Springer, and Stephen Tsai, three huge names in composites research in the United States.

Stanford has tremendous connections with industry in all its forms. Most of the Silicon Valley tech industry is populated with Stanford grads. And the tech industry has adopted composites for their structures wholeheartedly; a large number of the parts in some of your latest IT gadgets are composite.

UC Berkeley

Composites at UC Berkeley is headquartered in the Mechanical Engineering Department.[6] It is not as extensive as some of the other schools mentioned here, but Berkeley has an excellent mechanical engineering program, and it is a state school so in-state tuition is somewhat less expensive, and certainly cheaper than either Stanford or MIT. Their research is focused on biomechanics and biomaterials, medical polymers, and composites by additive manufacturing—intersection of materials research and computing. They also have excellent people in solid mechanics in the Mechanical Engineering Department; they are mostly in the fundamental science areas of composites and therefore much less focused on fabrication and manufacturing.

UCLA

UCLA is right in the heart of Southern California aerospace, and all the major aerospace companies have facilities there where they either assemble or manufacture some of the most advanced aircraft in the world. Those companies draw on talent from UCLA in the Mechanical & Aerospace Engineering Department[7] where the focus is mostly on mechanics of structural composites and composite manufacturing. In addition to that, and because the draw from the local aerospace industry is so great, UCLA has one of the more prominent campus extension programs in the United States. UCLA Extension[8] offers courses in composites manufacturing, composite structures, and several other topics related to composite materials. For lecturers and instructors, they hire the top talents in the industry to bring a practical, real-world experience to their students. They offer both classroom and online instruction. And on top of that, UCLA is ranked as the premier public university in the United States.

UT Austin

UT Austin[9] is most noted for their research in materials science, which also covers composite materials. They offer a broad range of courses for the undergrad and the graduate student in composite materials. These include composites design, mechanics of composites, and several courses in engineering materials science. They probably offer more for the undergraduate in composites, but UT Austin is a very good school and a very good springboard to moving on to advanced degrees or getting a job in composites directly out of school with a BS degree. They are well respected, and all the major oil companies draw on UT Austin for talent.

Rice University

Rice University houses their composites research in their Materials Science and NanoEngineering Department.[10] Their work spans from nanomaterials through the mesoscale, where the connection is made between microstructure at the nanoscale and macroscale materials properties, all the way up to development of novel polymers and polymer based composite materials with interesting optical, structural, and electronic properties. And Rice is another university in Texas that has excellent undergraduate as well as graduate education and is well respected in the industry. And while they don't have consortia with industry the way some others do, Rice is well connected to the oil and gas industry and is one of the leaders in the research into the use of carbon nanotubes to make carbon fiber. Their Carbon Hub launched in 2019 with $10 million from Shell and support from some other industrial companies in the carbon business.

Penn State

Penn State has been in the composites education and research business for some time. Their Applied Research Laboratory or Penn State ARL,[11] as it is known, was established by the Navy in 1945 and has been one of the premier Navy labs ever since. Their work in composite materials is broad and expansive and covers the watershed in technology development, process development, new materials, new process technologies, and computational analysis and design. They have an excellent composite materials lab, and both undergrad and graduate programs tailored to composites and composites engineering. Their Institute for Manufacturing and Sustainment Technologies is directly funded by the Office of Naval Research Manufacturing Technology (ManTech) Program and is the Navy's premier center for research and development in composite materials. That is partly why Penn State is ranked at the top university in materials science research according to the NSF.[12] Their Corporate Engagement Center has been active in partnering with industry for decades and has one of the best talent placement programs in the country for placing recent grads with BS degrees or advanced degrees in engineering and materials science.

University of Washington

At the University of Washington, composites engineering is headquartered in the Mechanical Engineering Department. Their most recent endeavor is to create the Advanced Composites Center (ACC)[13] to not only co-locate all the composites research that has been undertaken, but to address the needs of companies other than Boeing. Of course, the long-standing Boeing Advanced Research Center is still alive and well and will continue in partnership with the ACC, but the ACC is planning to be more expansive and will encompass all the manufacturing science and technology for aerospace composites under one umbrella. According to the director of the ACC, carbon fiber has come to the State of Washington in a big way, primarily because of the Boeing 787 and Boeing's advanced research into carbon fiber composites. He estimates that there are some 1400 companies and 130,000 jobs in the state involving advanced composites.[14] If you live in or near the state of Washington, and you want to work in the composites industry, UW is the place to be.

13

Final Thoughts

I certainly hope that you enjoyed this book, and more so that you got something out of it that you can use. Reading a book like this and not getting anything useful out of it is a waste of your time, and I don't want to waste your time, or mine.

I wrote the book with the idea that it would be an approachable treatment to a subject that seems unapproachable except to those who are already skilled in the art or are PhD engineers and have enough training to understand the sometimes-esoteric nature of these materials. The materials themselves are wonderful and endlessly fascinating. The stories about where they came from and who invented what, and even who stole someone else's idea is fascinating as well. The business has always been extraordinarily competitive and even sometimes cutthroat, but it is mostly practiced by people genuinely interested in doing good for the sake of others in society.

I also wrote the book in first person because I wanted it to be approachable. And, since I'm the approachable sort, I would love to hear what you think, and also to hear life stories because that's what this is all about, right? Working in composites is a very human endeavor, and each of the human beings involved in this industry has their own story and life experiences to share. That includes those who have been in it for some time, like myself, as well as those who are just starting out. Everyone has their own talents, aspirations, dreams, and goals and when they all join forces, great things can happen.

Almost anyone could have done the online research required to create this if they knew where to look and what to look for. That's why there are more than 100 references (notes) at the end of the book. Most of these are websites where you can go to get more information. Hopefully you can use this book as a guide to further your knowledge of composites and build on what's here to make a career out of this and become a composites engineer—if that is what you want to do. I intended this to be a bit evangelistic because I am somewhat of an evangelist about this subject and would love it if I knew that at least one or two people got enough out of reading this book that they were able to chart the course of their future in composite materials.

Wth that, I am going to sign off with a quote from Albert Einstein:

> **"Scientists investigate that which already is,
> engineers create that which has never been."**

Go become a composites engineer.

Chapter Notes

Chapter 1

1. BigRentz, "The History of Concrete," https://www.bigrentz.com/blog/the-history-of-concrete.

2. Danny Rosenberg, Serena Love, Emily Hubbard, and Florian Klimscha, "7,200 Years Old Constructions and Mudbrick Technology: The Evidence from Tel Tsaf, Jordan Valley, Israel," *PLOS ONE* 15, no. 1, e0227288. Accessed 22 January 2020.

3. For more info, see https://www.addcomposites.com/post/history-of-composites.

4. http://www.historyofmasks.net/mask-history/history-of-egyptian-masks/.

5. https://en.wikipedia.org/wiki/Leo_Baekeland.

6. https://en.wikipedia.org/wiki/Epoxy.

7. https://www.utoledo.edu/library/canaday/HTML_findingaids/MSS-222.html.

8. http://plastiquarian.com/?page_id=14253.

9. https://www.acs.org/content/acs/en/education/whatischemistry/landmarks/carbonfibers.html.

10. https://en.wikipedia.org/wiki/Polyacrylonitrile.

11. H. Finkentscher and C. Heuck, German patent for polyacrylonitrile (PAN), IG Farben, DE Patent 654989, Verfahren zur Herstellung von Polymerisationprodukten, Anmeldetag 18.2.1930.

12. https://acs.confex.com/acs/cerm09/webprogram/Paper71257.html.

Chapter 2

1. https://people.wou.edu/~courtna/ch412/perhist.htm.

2. Highly suggested reading: Sam Kean, *The Disappearing Spoon* (New York: Little, Brown, 2010). This is a very engaging book about the heyday of element "discovery" and the intense and sometimes brutal competition among scientists to be the "discoverer" of an element.

3. "Copper History," Rameria.com, archived from the original 17 September 2008. Retrieved 12 September 2008.

4. "History of Carbon," https://www.caer.uky.edu/carbon/history/carbonhistory.shtml, archived from the original 1 November 2012. Retrieved 10 January 2013.

5. https://en.wikipedia.org/wiki/Timeline_of_chemical_element_discoveries.

6. https://en.wikipedia.org/wiki/Silicon.

7. https://www.aps.org/publications/apsnews/200605/history.cfm.

8. Niels Bohr, "On the Constitution of Atoms and Molecules, Part II Systems Containing Only a Single Nucleus," *Philosophical Magazine* 26, no. 153 (1913): 476–502. http://web.ihep.su/dbserv/compas/src/bohr13b/eng.pdf.

9. Chemistry Operations, "Carbon," 15 December 2003, Los Alamos National Laboratory, archived from the original 13 September 2008.

10. U.S. Patent Number 2230272: Method of Producing Glass Fibers, which can be found in the Wikipedia article on Games Slayter, https://en.wikipedia.org/wiki/Games_Slayter.

11. http://www.chemistryexplained.com/Ge-Hy/Glass.html.

12. https://en.wikipedia.org/wiki/Epoxy.

13. https://en.wikipedia.org/wiki/Bisphenol_A.

Chapter 3

1. F. Rozploch, J. Patyk, J. Stanowski, "Graphenes Bonding Forces in Graphite," *Acta Physica Polonica A* 112, no. 3 (2007): 557–563.
2. Abu Yaya, "Layered Nanomaterials—A Review," *Global Journal of Engineering Design and Technology* 1 (2012): 32–41.
3. https://www.acs.org/content/acs/en/education/whatischemistry/landmarks/carbonfibers.html.
4. See Wikipedia for an explanation of Faraday cage—https://en.wikipedia.org/wiki/Faraday_cage
5. https://en.wikipedia.org/wiki/Acrylonitrile.
6. https://www.compositesworld.com/articles/the-making-of-carbon-fiber.
7. https://polser.com/en/frp/fiber glass-types.
8. https://en.wikipedia.org/wiki/Glass_fiber.
9. "What Is Kevlar," DuPont, https://web.archive.org/web/20070320005408/http:/www.dupont.com/kevlar/whatiskevlar.html, archived from the original 20 March 2007.
10. https://www.teijinaramid.com/en/products/twaron/.
11. https://en.wikipedia.org/wiki/Vectran.
12. https://www.compositesworld.com/articles/boron-fiber-the-original-high-performance-fiber.
13. https://en.wikipedia.org/wiki/Moissanite.
14. G. Acheson, U.S. Patent 492,767, "Production of artificial crystalline carbonaceous material," 1893.
15. https://en.wikipedia.org/wiki/Silicon_carbide_fibers.
16. Seishi Yajima, Josaburo Hayashi, and Mamoru Omori, U.S. patent 4,052,430, "Method for producing organosilicon high molecular weight compounds having silicon and carbon as main skeleton components and said organosilicon high molecular weight compounds," issued 4 October 1977.

Chapter 4

1. https://www.plenco.com/phenolic-novolac-resol-resins.htm.
2. https://worldiscoveries.ca/technologies/formaldehyde-free-phenolic-bio-resins/.
3. https://www.ncbi.nlm.nih.gov/pmc/articles/PMC7693354/.
4. https://pubs.rsc.org/en/content/articlelanding/2014/ra/c4ra04458d/unauth.
5. https://polymerdatabase.com/polymer%20classes/Unsaturated%20Polyester%20type.html
6. https://www.sciencedirect.com/science/article/pii/B9780127639529500058.
7. https://polymerinnovationblog.com/-thermoset-cure-chemistry-part-3-epoxy-curing-agents/.
8. "Victrex Celebrates 40 Years of PEEK Success," Victrex, retrieved 1 November 2021.

Chapter 5

1. Christopher H. Childress, "Determination of Thermoplastic Composite Crystallization Process Limits for Dynamic and Isothermal Cooling Processes," *Boeing Technical Journal*, 2019.
2. https://www.compositesworld.com/articles/tooling.
3. https://www.aniwaa.com/buyers-guide/3d-printers/largest-3d-printers/.
4. https://www.sciaky.com/largest-metal-3d-printer-available.
5. https://www.compositesworld.com/articles/the-evolution-of-infusion.
6. K.N. Kendall, C.D. Rudd, M.J. Owen, and V. Middleton, "Characterization of the Resin Transfer Moulding Process," *Composites Manufacturing* 3, no. 4 (1 January 1992): 235–249.
7. https://www.compositesworld.com/articles/autoclave-quality-outside-the-autoclave.
8. https://www.compositesworld.com/articles/sqrtm-enables-net-shape-parts.
9. *Delaware Composites Design Encyclopedia: Processing and Fabrication Technology*, Volume 3 (Lancaster, PA: Technomic Publishing Company, 1990).

Chapter 6

1. *Delaware Composites Design Encyclopedia: Mechanical Behavior and Properties of Composite Materials*, Volume 1 (Lancaster, PA: Technomic Publishing Company, 1990).
2. https://www.espcomposites.com/software/download.html.
3. R.M. Christensen and E. Zywicz, "A Three-Dimensional Constitutive Theory for Fiber Composite Laminated Media," *Journal of Applied Mechanics* 57, no. 4 (1 December 1990): 948–955.

Chapter 7

1. *Delaware Composites Design Encyclopedia: Design Studies*, Volume 5 (Lancaster, PA: Technomic Publishing Company, 1990).
2. https://www.youtube.com/watch?v=4DKkueqcKmQ.

Chapter 8

1. Tianliang Qin, Libin Zhao, and Jianyu Zhang, "Fastener Effects on Mechanical Behaviors of Double-Lap Composite Joints," *Composite Structures* 100 (2013): 413–423.
2. https://en.wikipedia.org/wiki/Delamination.
3. M.J. Hinton, A.S. Kaddour, and P.D. Soden "Predicting Failure in Fibre Composites: Lessons Learned from the World-Wide Failure Exercise," British Crown Copyright 2000. Published with the permission of the Defence Evaluation and Research Agency on behalf of the Controller of HMSO.

Chapter 9

1. https://en.wikipedia.org/wiki/Nastran.
2. https://en.wikipedia.org/wiki/Femap.
3. https://en.wikipedia.org/wiki/SolidWorks.

Chapter 10

1. https://en.wikipedia.org/wiki/Metal_matrix_composite.

2. https://www.tms.org/pubs/journals/jom/0104/rawal-0104.html.
3. https://en.wikipedia.org/wiki/Ceramic_matrix_composite.

Chapter 11

1. https://www.compositesworld.com/articles/the-making-of-glass-fiber.
2. https://www.businesswire.com/news/home/20211013005589/en/The-Worldwide-Glass-Fiber-Industry-is-Expected-to-Reach-10.6-Billion-by-2026---ResearchAndMarkets.com.
3. https://www.teijincarbon.com/about-us/company-history.
4. https://www.toray.com/global/aboutus/history/.
5. https://www.compositesworld.com/articles/carbon-fiber-in-automotive-at-a-dead-end.
6. https://www.nytimes.com/1988/05/11/business/company-news-akzo-dupont-deal-ends-11-year-fight.html.
7. DuPont, Kevlar Technical Guide 0319.
8. https://news.nike.com/news/nike-basketball-s-superhero-elite-series-2-0-rises-above-the-rest.
9. https://www.thedyneemaproject.com/en_GB/the-fabrics/dyneema.html.
10. https://www.globenewswire.com/news-release/2021/06/14/2246878/0/en/Unsaturated-Polyester-Resin-Market-to-Touch-USD-16-965-7-Million-by-2027-Rising-Robustness-and-Durability-of-Product-to-Boost-Market-Growth-Says-Fortune-Business-Insights.html.
11. https://en.wikipedia.org/wiki/Ineos.
12. http://www.sinopolymer.cn/Who_We_Are/About_Sino_Polymer/.
13. https://www.mordorintelligence.com/industry-reports/vinyl-ester-market.
14. https://en.wikipedia.org/wiki/Hexion.
15. https://www.huntsman.com/about/our-history.
16. https://www.grandviewresearch.com/industry-analysis/epoxy-resins-market.
17. https://www.globenewswire.com/news-release/2021/10/21/2318848/0/en/Global-Epoxy-Composites-Market-is-Estimated-to-Observe-USD-47-17-billion-by-2028-Fior-Markets.html.

18. https://space.stackexchange.com/questions/22003/how-can-phenolic-resin-handle-rocket-engine-nozzle-temperatures.

19. https://plenco.com/plenco-company-history.htm.

20. https://www.marketsandmarkets.com/Market-Reports/phenolic-resin-market-177821389.html.

21. https://www.compositesworld.com/articles/2021-cw-top-shops-highlights-strengths-of-top-composites-facilities.

22. https://www.marketsandmarkets.com/Market-Reports/composite-market-200051282.html.

Chapter 12

1. https://www.ccm.udel.edu/industry/.

2. https://www.mccormick.northwestern.edu/materials-science/research/areas-of-research/composites.html.

3. https://necstlab.mit.edu/.

4. https://mse.gatech.edu/research-area/composites.

5. https://sacl.stanford.edu/.

6. https://me.berkeley.edu/research-areas-and-major-fields/materials/.

7. https://www.mae.ucla.edu/mechanics-of-structural-composites/.

8. https://www.uclaextension.edu/.

9. https://tmi.utexas.edu/.

10. https://msne.rice.edu/.

11. https://www.arl.psu.edu/.

12. https://corporateengagement.psu.edu/focus-areas/material-science/.

13. https://depts.washington.edu/uwacc/node/1.

14. https://www.me.washington.edu/news/article/2021-03-22/how-washington-became-global-epicenter-advanced-carbon-fiber.

Bibliography

Bacon, Roger. U.S. Patent # 2957756, "Filamentary graphite and method for producing the same," issued 25 October 1960.

Bohr, Niels. "On the Constitution of Atoms and Molecules, Part II Systems Containing Only a Single Nucleus." *Philosophical Magazine* 26, no. 153 (1913): 476–502. PDF.

Bowden, Mary Ellen. "Leo Baekeland." Chemical Achievers: The Human Face of the Chemical Sciences. *Philadelphia: Chemical Heritage Foundation, 1997.*

Chawla, Krishan K. *Composite Materials Science and Engineering*, 4th ed. Cham: Springer, 2014. This is a very widely used textbook on composites that—in the fourth edition—offers a fully integrated and completely up to date treatise on composite materials. This is one of the classic college texts that is used in quite a few graduate and upper level classes in composites in some of the better universities. Great reference book to have on your bookshelf.

Childress, Christopher H. "Determination of Thermoplastic Composite Crystallization Process Limits for Dynamic and Isothermal Cooling Processes." *Boeing Technical Journal*, 2019.

Christensen, R.M., and E. Zywicz. "A Three-Dimensional Constitutive Theory for Fiber Composite Laminated Media." *Journal of Applied Mechanics* 57, no. 4 (1 December 1990): 948–955. One of the seminal papers in development of the means to analyze thick composites using a Finite Element program.

Cook, J. Gorgon. *Handbook of Textile Fibres: Man-Made Fibres*. Cambridge: Woodhead, 1984.

"Copper History." Rameria.com, http://www.rameria.com/inglese/history.html.

Delaware Composites Design Encyclopedia. Lancaster, PA: Technomic Publishing Company, 1990. This encyclopedia is exactly what it says it is, an encyclopedia of composite materials. This six-volume set includes *Mechanical Behavior and Properties of Composites*, *Micromechanical Materials Modeling, Processing and Fabrication Technology, Failure Analysis, Design Studies*, and *Test Methods for Composites*. Well worth owning at least a digital set.

Dupont. Kevlar Technical Guide 0319.

Finkentscher, H., and C. Heuck. German Patent for Polyacrylonitrile (PAN). IG Farben, DE Patent 654989, Verfahren zur Herstellung von Polymerisationprodukten, Anmeldetag 18.2.1930.

Gordin, Michael D. *A Well-Ordered Thing: Dmitrii Mendeleev and the Shadow of the Periodic Table*. New York: Basic Books, 2004.

Gupta, A.K., D.K. Paliwal, and P. Bajaj. "Melting Behavior of Acrylonitrile Polymers." *Journal of Applied Polymer Science* 70, no. 13 (1998): 2703–2709.

Halpin, John C. *Primer on Composite Materials Analysis,* 2d ed., rev. Boca Raton: CRC Press, 1992. This is a great starter book on the mechanics of composites written in a manner that is understandable.

"High Performance Carbon Fibers." *National Historic Chemical Landmarks,* American Chemical Society, 2003.

Hinton, M.J., A.S. Kaddour, and P.D. Soden. "Predicting Failure in Fibre Composites: Lessons Learned from the World-Wide Failure Exercise." British Crown Copyright 2000. Published with the permission of the Defence Evaluation and Research Agency on behalf of the Controller of HMSO.

"History of Carbon." https://web.archive.org/web/20121101085829/http:/www.caer.uky.edu/carbon/history/carbonhistory.shtml.

https://en.wikipedia.org/wiki/Timeline_of_chemical_element_discoveries.

https://plenco.com/plenco-company-history.htm.

https://www.grandviewresearch.com/industry-analysis/epoxy-resins-market.

Jones, Robert M. Mechanics of Composite Materials, 2d ed. Boca Raton: CRC Press, 1999. This is something of a bible in the mechanics of composites world. Robert Jones is one of the best and most knowledgeable in the business. This was written when he was senior technical and Wright Patterson AFB, leading the development of some of the most advanced composite aircraft ever built.

Kean, Sam. The Disappearing Spoon. New York: Little, Brown, 2010. This is a very engaging book about the heyday of element "discovery" and the intense and sometimes brutal competition among scientists to be the "discoverer" of an element. Well worth owning a copy.

May, Clayton. Epoxy Resins: Chemistry and Technology, 2d ed. Boca Raton: CRC Press, 2018.

Mendeleeva, Maria. D.I. Mendeleev's Archive: Autobiographical Writings. Collection of Documents. Volume 1 // Biographical notes about D. I. Mendeleev (written by me—D. Mendeleev), p. 13. Leningrad: D. I. Mendeleev's Museum-Archive, 1951. Two hundred and seven pages (in Russian).

Miodownik, Mark. Stuff Matters. Boston: Houghton Mifflin Harcourt, 2013. This is another book that is well worth reading. The author is a material scientist and he explains why the periodic table of the elements is so important to understand. That's where we get all of the stuff that we use every day.

Qin, Tianliang, Libin Zhao, and Jianyu Zhang. "Fastener Effects on Mechanical Behaviors of Double-Lap Composite Joints." Composite Structures 100 (2013): 413–423.

Reddy, J.N. Mechanics of Laminated Composite Plates and Shells—Theory and Analysis, 2d ed. Boca Raton: CRC Press, 2004.

Rosenberg, Danny, Serena Love, Emily Hubbard, and Florian Klimscha. "7,200 Years Old Constructions and Mudbrick Technology: The Evidence from Tel Tsaf, Jordan Valley, Israel." PLOS ONE 15, no. 1: e0227288. Accessed 22 January 2020.

Rozploch, F., J. Patyk, and J. Stanowski. "Graphenes Bonding Forces in Graphite." Acta Physica Polonica A 112, no. 3 (2007): 557–563.

Sathishkumar, T.P., S. Satheeshkumar, and J. Naveen. (July 2014). "Glass Fiber-Reinforced Polymer Composites—A Review." Journal of Reinforced Plastics and Composites 33, no. 13 (July 2014): 1258–1275.

Science Direct. "Inorganic and Composite Fibers." www.sciencedirect.com.

Slayter Patent for Glass Wool. Application 1933, granted 1938.

Tsai, Stephen W. Composites Design, 3d ed. Dayton: Think Composites, 1987. This is another bible in composites—Dr. Tsai is one of the most respected names in composites design. He is currently an emeritus research professor in Stanford's Aeronautics and Astronautics Department.

Wanberg, John. Composite Materials—Step-by-Step Projects. Sillwater, MN: Wolfgang Publications, 2014. Professor John Wanberg wrote three other step-by-step composites how-to books—Composite Materials: Fabrication Handbooks, Volumes 1–3—that have become standard fare in and are some of the best of the how-to books written about composites and how to make things using these wonderful materials.

Yaya, Abu. (2012). "Layered Nanomaterials—A Review." Global Journal of Engineering Design and Technology 1 (2012): 32–41.

Index